algebra for all green level

elizabeth warren PhD

Series Consultants
James Burnett MEd
Calvin Irons PhD

ORIGO
EDUCATION

About the Author

Elizabeth Warren has been involved in mathematics education for more than 30 years. During this time, she has actively worked in both the university and school levels, and engaged both elementary and secondary school teachers in professional development activities. She is currently conducting research into patterns and algebra.

Algebra for All, Green Level

Copyright 2007 ORIGO Education
Author: Elizabeth Warren PhD
Series Consultants: James Burnett MEd and Calvin Irons PhD

Warren, Elizabeth.
Algebra for All: green level.

ISBN 1 921023 04 X.
1. Algebra - Problems, exercises, etc. - Juvenile
literature. I. Title.
512

For more information, contact
North America
Tel. 1-888-ORIGO-01 or 1-888-674-4601
Fax 1-888-674-4604
sales@origomath.com
www.origomath.com

UK/Australasia
info@origo.com.au
www.origo.com.au

ISBN: 978 1 921023 04 0

10 9 8 7 6 5 4 3 2

Printed using a waterless printing process that reduces toxins in our waterways and minimizes overall water use.

INTRODUCTION

What is algebra?

Algebraic thinking commences as soon as students identify consistent change and begin to make generalizations. Their first generalizations relate to real-world experiences. For example, a child may notice a relationship between her age and the age of her older brother. In the example below, Ali has noted that her brother Brent is always 2 years older than her.

Ali's age	Brent's age
8	10
9	11
10	12
11	13

Over time these generalizations extend to more abstract situations involving symbolic notation that includes numbers. The above relationship can be generalized using the following symbolic notation.

$$Ali + 2 = Brent \qquad A + 2 = B$$

Algebraic thinking uses different symbolic representations, such as unknowns and variables, with numbers to explore, model, and solve problems that relate to change and describe generalizations. The symbol system used to describe generalizations is formally known as algebra. Following the example above, Ali wonders how old she will be when Brent is 21 years old. We can solve a problem such as this by "backtracking" the generalization ($A = 21 - 2$) or using the balance method of subtracting 2 from both sides the equation ($Ali = Brent - 2$).

Why algebra?

Identifying patterns and making generalizations are fundamental to all mathematics, so it is essential that students engage in activities involving algebra. Many practical uses for algebra lie hidden under the surface of an increasingly electronic world — specific rules are used to determine telephone charges, track bank accounts and generate statements, describe data represented in graphs, and encrypt messages to make the Internet secure. Algebraic thinking is more overt when we create rules for spreadsheets or simply use addition to solve a subtraction problem.

Algebra involves the generalizations that are made regarding the relationships between variables in the symbol system of mathematics.

What are the "big ideas"?

The lessons in the *Algebra for All* series aim to develop the "big ideas" of early algebra while supporting thinking, reasoning, and working mathematically. These ideas of equivalence and equations, patterns and functions, properties, and representations are inherent in all modern curricula and are summarized in the following paragraphs.

Equivalence and Equations

The most important ideas about equivalence and equations that students need to understand are:

- "Equals" indicates equivalent sets rather than a place to write an answer
- Simple real-world problems with unknowns can be represented as equations
- Equations remain true if a consistent change occurs to each side (the balance strategy)
- Unknowns can be found by using the balance strategy.

Patterns and Functions

This idea focuses on mathematics as "change". Change occurs when one or more operation is used. For example, the price of an item bought on the Internet changes when a freight charge is added. It is important for students to understand that:

- Operations almost always change an original number to a new number
- Simple real-world problems with variables can be represented as "change situations"
- "Backtracking" reverses a change and can be used to solve unknowns.

Properties

Students will discover a variety of arithmetic properties as they explore number, such as:

- The commutative law and the associative law exist for addition and multiplication but not for subtraction and division
- Addition and subtraction are inverse operations, as are multiplication and division
- Adding or subtracting zero and multiplying or dividing by 1 leaves the original number unchanged
- In certain circumstances, multiplication and division distribute over addition and subtraction.

Representations

Different representations deepen our understanding of real-world problems and help us identify trends and find solutions. This idea focuses on creating and interpreting a variety of representations to solve real-world problems. The main representations that are developed in this series include graphs, tables of values, drawings, equations, and everyday language.

INTRODUCTION

About the series

Each of the six *Algebra for All* books features 4 chapters that focus separately on the "big ideas" of early algebra — Equivalence and Equations, Patterns and Functions, Properties, and Representations. Each chapter provides a carefully structured sequence of lessons. This sequence extends across the series so that students have the opportunity to develop their understanding of algebra over a number of years.

About the lessons

Each lesson is described over 2 pages. The left-hand page describes the lesson itself, including the aim of the lesson, materials that are required, clear step-by-step instructions, and a reflection. These notes also provide specific questions that teachers can ask students, and subsequent examples of student responses. The right-hand page supplies a reproducible blackline master to accompany the lesson. The answers for all blackline masters can be found on pages 66-73.

Subtitles indicate lesson content.

Reproducible blackline masters contain fun and engaging activities for practice.

Simple and concise step-by-step instructions are provided for each lesson.

The aim of the lesson is clearly stated.

Materials required for the lesson are listed in order of use.

A reflection suggests activities and class discussions to consolidate learning.

Suggested questions and examples of student responses are included in each lesson.

Side tabs indicate the lesson number in each chapter.

Footer notes indicate the corresponding chapter for each lesson.

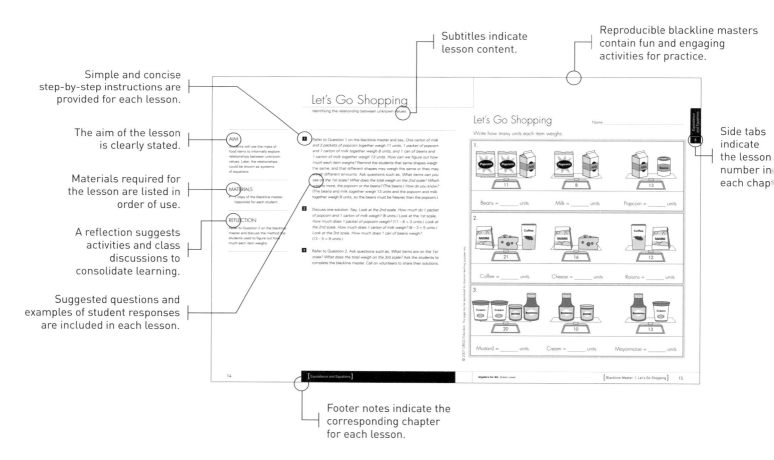

Assessment

Students' thinking is often best gauged by the conversations that occur during classroom discussions. Listen to your students and make notes about their thinking. You may decide to use the rubric below to assess students' mathematical proficiency in the tasks for each lesson. Study the criteria, then assess and record each student's understanding on a copy of the Assessment Summary provided on page 74. Although the summary lists every lesson in this book, it is not necessary to assess students for all lessons.

A	The student fully accomplishes the purpose of the task. Full understanding of the central mathematical ideas is demonstrated. The student is able to communicate his/her thinking and reasoning.
B	The student substantially accomplishes the purpose of the task. An essential understanding of the central mathematical ideas is demonstrated. The student is generally able to communicate his/her thinking and reasoning.
C	The student partially accomplishes the purpose of the task. A partial or limited understanding of the central mathematical ideas is demonstrated and/or the student is unable to communicate his/her thinking and reasoning.
D	The student is not able to accomplish the purpose of the task. Little or no understanding of the central mathematical ideas is demonstrated and/or the student's communication of his/her thinking and reasoning is vague or incomplete.

Simple Stories

Writing simple addition stories with unknowns

AIM

Students will explore and write simple addition stories with unknowns for real-world contexts.

MATERIALS

- 1 copy of the blackline master (opposite) for each student

REFLECTION

Ask, *When we write a word problem for addition with unknowns, what do we need to think about first?* (Context.) *What do we think about next?* (Numbers in that context.) *What words can we use to describe the addition? What words can we use to describe the equal amounts?* (Same as.)

1 Say, *If Rhian saves $5 more, she will have the same amount of money as Shauna. Shauna has $15. How much money does Rhian have in her piggy bank? How can we show this?* Elicit several suggestions then draw a model of the story on the board, representing the unknown amount as a square. Ask, *How can we write this as an equation?* Invite suggestions, then write $\square + 5 = 15$ on the board.

2 Write $\square + 3 = 8$ on the board. Ask, *How many are on the left side?* ($\square + 3$) *How many are on the right?* (8) *What story can we write about this equation?* Encourage the students to identify different contexts as a story basis, for example, ducks in 2 ponds, plants on 2 steps, or students in 2 classrooms. Illustrate their suggestions on the board then encourage them to write 2 stories using different contexts. Point out the key components of their stories, for example, 2 ponds (2 steps or 2 classrooms) with "$\square + 3$" linked to one side and "8" linked to the other, and a statement to describe the amounts as the same.

3 Ask the students to complete the blackline master. Invite volunteers to share their stories. Discuss the different contexts they used and how they decided what the numbers and unknowns represented. Highlight the different words they used for addition in those contexts, and the equivalence statements they wrote.

Simple Stories

Name _____

1. One tree has 10 birds. Another tree has some birds and 3 more join them.
   Now the number of birds in each tree is the same.

a. Write numbers in this balance picture to match the story.

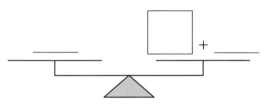

b. Write the equation.

_____ = ⬜ + _____

2. a. Write a story to match ⬜? + 4 = 11. _____

   _____

   b. Draw a picture to show your story.

   c. Write numbers in this balance picture to match your story.

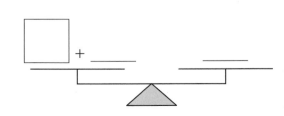

More Simple Stories

Writing simple subtraction stories with unknowns

AIM

Students will explore and write simple subtraction stories with unknowns for real-world contexts. Although the students may want to figure out the unknown numbers, this is not the main aim of the activity.

MATERIALS

- 1 copy of the blackline master (opposite) for each student

REFLECTION

Invite individuals to share the stories they wrote on the blackline master. Reinforce the idea that both sides of the equation are the same. One side is the unknown starting number (total) and a known amount to be subtracted (part of the total), and the other side is the known number left (part of the total).

1 Say, *Jean has a full box of chocolates. She gives 9 chocolates away. There are 16 chocolates left in the box. How can we draw this as a balance picture?* Elicit several suggestions then illustrate the story on the board, as shown below.

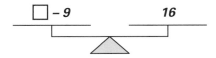

Ask, *What equation can we write?* Invite suggestions, then write ☐ **– 9 = 16** on the board.

2 Ask the students to identify and draw contexts for subtraction stories, for example, people in a full room, and a full jar of cookies. Share an example of a subtraction story. Say, *The cookie jar is full. If 3 cookies are eaten there will be 8 cookies left in the jar. How can we show this story on a balance scale?* Elicit responses then draw a balance scale with "8" on one side and "☐ – 3" on the other side.

3 Ask the students to complete the blackline master. Call on volunteers to share their stories. Discuss with the students the different contexts they used and how they decided what the numbers and unknowns represented. Highlight the different ways they described the subtraction (spent, ate, left) and the equivalence statements they wrote.

More Simple Stories

Name _____

1. Six frogs jump out of a pond. Ten frogs stay in the pond. How many are there in total?

a. Write numbers in this balance picture to match the story.

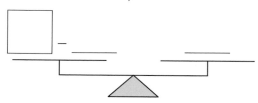

b. Write the equation.

_____ = [ ] − _____

2. a. Write a story to match [?] − 2 = 8. _____

    _____

    _____

    b. Draw a picture to show your story.

    c. Write numbers in this balance picture to match your story.

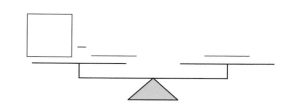

Greater Than, Less Than

Working with > and < in multiplication situations

AIM

Students will use > and < in multiplication situations.

MATERIALS

- 1 sheet of paper for each pair of students

- 1 copy of the blackline master (opposite) for each student

REFLECTION

Refer to Question 1 on the blackline master and ask, *How did you figure out values for the unknowns? How can you figure out the greatest value for each unknown?*

1 Ask students to work in pairs. Provide pairs with a sheet of paper. Say, *Draw rows of apples with the same number in each row. You cannot draw a total that is more than 50 apples.* After the students draw their arrays, invite two pairs of students to show their array pictures and describe what is the same and different about the pictures. Encourage members of the class to compare the total number of apples in the two arrays and explain their thinking.

2 Draw a picture of a balanced scale and another of an unbalanced scale. Ask the students to place their two array pictures on the correct scale and write the related number sentence underneath as shown in the two examples below.

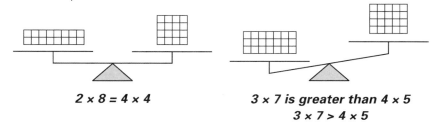

2 x 8 = 4 x 4   **3 x 7 is greater than 4 x 5**
                                    3 x 7 > 4 x 5

Repeat the discussion for other pairs of array pictures.

3 On the board, draw the diagram shown below.

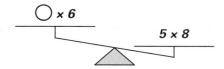

Point to the circle and ask, *What whole numbers can we write here to make this true? What is the greatest whole number that we can use? What is the least whole number?* Write $\bigcirc$ **x 6 ____ 5 x 8** and **5 x 8 ____ $\bigcirc$ x 6** on the board and ask, *What symbols will make these match the balance scale?*

4 Ask the students to complete the blackline master. Call on volunteers to share their answers.

Greater Than, Less Than

1. Draw arrays to make these true. Then complete the matching number sentences.

a.

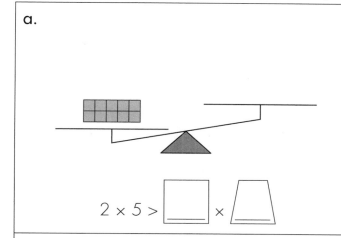

$2 \times 5 >$ ⬜ $\times$ ⬜

b.

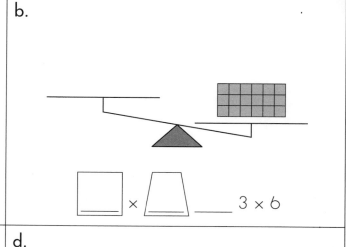

⬜ $\times$ ⬜ ____ 3×6

c.

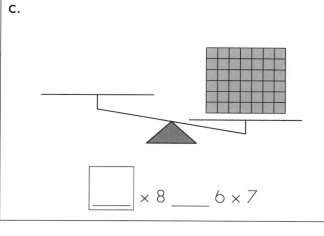

⬜ $\times 8$ ____ 6×7

d.

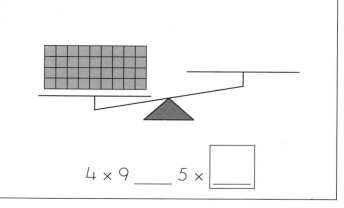

4×9 ____ $5 \times$ ⬜

2. Write **<, >**, or **=** to make these true.

a. 6×2 ____ 3×4 b. 5×7 ____ 4×9

c. 8×4 ____ 3×10 d. 4×12 ____ 5×9

e. 9×7 ____ 6×11 f. 4×8 ____ 2×6

g. 8×0 ____ 0×12 h. 9×6 ____ 8×7

Looking for Clues

Determining the relationship between collections of objects

AIM

Students will use visual clues to determine relationships between collections of objects.

MATERIALS

- 1 copy of the blackline master (opposite) for each student

REFLECTION

Refer to the blackline master and ask, *How do the clues help us solve these types of problems?*

1 Refer to Question 1 on the blackline master. Call on volunteers to read the clues. Ask, *How many spheres balance 3 cubes?* (2) *How many cubes balance 1 cylinder?* (2) *How can we balance 2 spheres and 2 cubes?* (Add either 1 cylinder or 2 cubes to the left side.) Encourage the students to share and justify their solutions.

2 Refer to Question 2. Ask, *How many cylinders balance 2 rectangular prisms?* (4) *So how many cylinders balance 1 rectangular prism?* (2) *If we substitute cylinders for the prisms in the 2nd clue, what will the picture show?* (1 cone + 6 cylinders = 2 cones.) *Which item can we take away from both sides and keep the scale balanced?* (1 cone.) *What will be left on the scale?* (6 cylinders on one side and 1 cone on the other side.) *So which items do we add to the unbalanced scale to make it balance?* (Add 3 cylinders to the left side.) Encourage the students to complete Question 2b. Allow time for them to share their answers.

Looking for Clues

Name _____

1.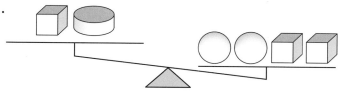

a. Look at the clues. What do you need to add to the left side of the scale above to make it balance?

_____

b. Write how you figured it out.

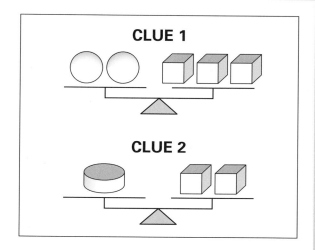

CLUE 1

CLUE 2

_____

_____

_____

2.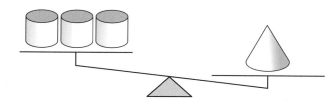

a. Look at the clues. What do you need to add to the left side of the scale above to make it balance?

_____

b. Write how you figured it out.

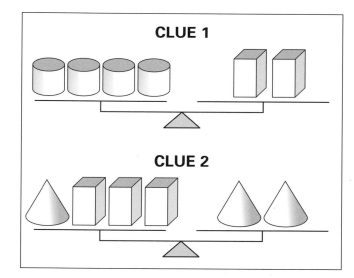

CLUE 1

CLUE 2

_____

_____

_____

Let's Go Shopping

Identifying the relationship between unknown values

AIM

Students will use the mass of food items to informally explore relationships between unknown values. Later, the relationships could be shown as systems of equations.

MATERIALS

- 1 copy of the blackline master (opposite) for each student

REFLECTION

Refer to Question 2 on the blackline master and discuss the method the students used to figure out how much each item weighs.

1 Refer to Question 1 on the blackline master and say, *One carton of milk and 2 packets of popcorn together weigh 11 units, 1 packet of popcorn and 1 carton of milk together weigh 8 units, and 1 can of beans and 1 carton of milk together weigh 13 units. How can we figure out how much each item weighs?* Remind the students that same shapes weigh the same, and that different shapes may weigh the same or they may weigh different amounts. Ask questions such as, *What items can you see on the 1st scale? What does the total weigh on the 2nd scale? Which weighs more, the popcorn or the beans?* (The beans.) *How do you know?* (The beans and milk together weigh 13 units and the popcorn and milk together weigh 8 units, so the beans must be heavier than the popcorn.)

2 Discuss one solution. Say, *Look at the 2nd scale. How much do 1 packet of popcorn and 1 carton of milk weigh?* (8 units.) Look at the 1st scale. *How much does 1 packet of popcorn weigh?* (11 – 8 = 3 units.) *Look at the 2nd scale. How much does 1 carton of milk weigh?* (8 – 3 = 5 units.) *Look at the 3rd scale. How much does 1 can of beans weigh?* (13 – 5 = 8 units.)

3 Refer to Question 2. Ask questions such as, *What items are on the 1st scale? What does the total weigh on the 3rd scale?* Ask the students to complete the blackline master. Call on volunteers to share their solutions.

Let's Go Shopping

Name _____

Write how many units each item weighs.

1.

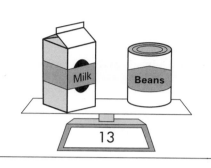

Beans = _____ units   Milk = _____ units   Popcorn = _____ units

2.

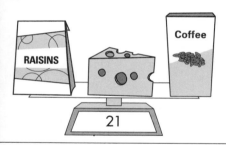

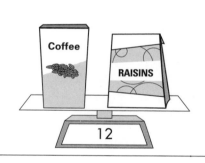

Coffee = _____ units   Cheese = _____ units   Raisins = _____ units

3.

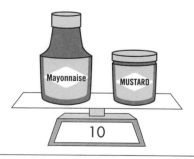

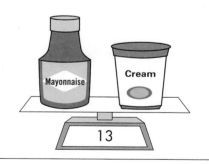

Mustard = _____ units   Cream = _____ units   Mayonnaise = _____ units

Counter Balance

Multiplying both sides of simple addition equations by the same amount

AIM

Students will conclude that if both sides of an addition equation are multiplied by the same number, the equation remains balanced.

MATERIALS

- 1 set of balance scales
- Magnetic counters (or standard counters and Blu-Tack)
- Small plastic bags
- Connecting cubes
- 1 copy of the blackline master (opposite) for each student

TEACHING NOTE

If the students are not familiar with the use of parentheses ( ), use the bags to help explain that the numbers are added together before they are doubled or tripled.

REFLECTION

Discuss the examples on the blackline master and then ask, *If we multiply one side of an equation by an unknown, what must we do to keep the equation balanced?* (Multiply the other side by the same amount.)

1 Ask volunteers to use red and blue connecting cubes to show the different combinations that total 5 cubes. Each combination should be placed in a plastic bag. Place 2 bags on the scale as shown below and ask, *What number sentence can we write?* Write the equation on the board and ask, *Why is this an equation?* (It has an equals symbol and is balanced.)

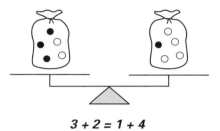

3 + 2 = 1 + 4

2 Invite volunteers to put cubes in bags to match each side of the scale and place the bags on the scale as shown below.

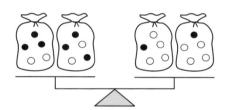

2 bags of (3 + 2) = 2 bags of (1 + 4)
2 × (3 + 2) = 2 × (1 + 4)

Ask, *What did we do to each side of the scale?* (Doubled the number of bags.) *Is the scale still balanced? Why? How can we write this as an equation?* Elicit several responses and then write the new equation (shown above) on the board. Repeat for tripling the number of bags on each side of the scale. If time allows, repeat for another pair of bags.

3 Have the students complete the blackline master. Call on volunteers to share their answers.

Counter Balance

1. Keep the picture balanced by drawing **double** the number of counters on each side.
    Then complete the new equation.

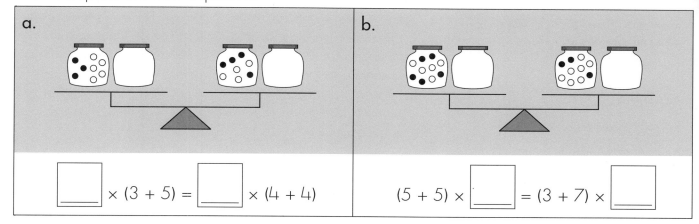

a. $\boxed{} \times (3 + 5) = \boxed{} \times (4 + 4)$

b. $(5 + 5) \times \boxed{} = (3 + 7) \times \boxed{}$

2. Keep the picture balanced by drawing **triple** the number of counters on each side.
    Then complete the new equation.

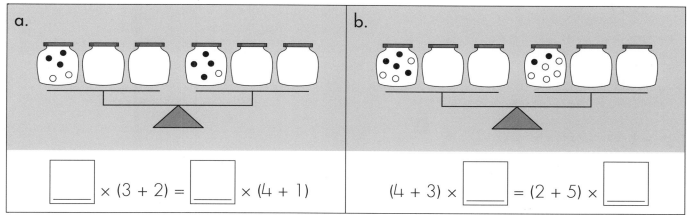

a. $\boxed{} \times (3 + 2) = \boxed{} \times (4 + 1)$

b. $(4 + 3) \times \boxed{} = (2 + 5) \times \boxed{}$

3. Complete the matching picture and equation.

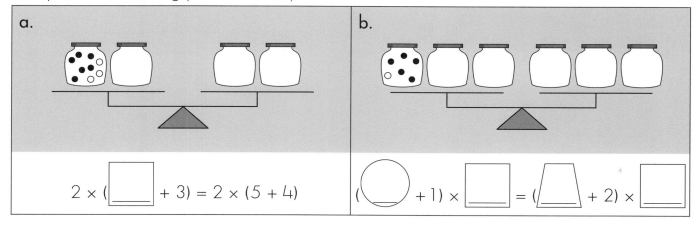

a. $2 \times (\boxed{} + 3) = 2 \times (5 + 4)$

b. $(\bigcirc + 1) \times \boxed{} = (\diagup\diagdown + 2) \times \boxed{}$

Finding Missing Values

Solving equations with one unknown value

AIM

Students will use balance to solve problems with one unknown.

MATERIALS

- 1 set of balance scales
- Identical lightweight paper bags
- Connecting cubes
- 1 copy of the blackline master (opposite) for each student

REFLECTION

Refer to the examples on the blackline master. Ask, *How did you figure out the value of the unknown in an addition (multiplication) equation?* (Subtract/divide.) *What can we do to figure out an unknown in a subtraction (division) equation? How can we check that our answer is correct?* (Substitute the solution in the equation.)

1 Review the concept of using balance to solve for unknowns in simple addition and subtraction equations. Write $\square + 7 = 12$ on the board. Model the equation with the balance scale and cubes. Represent the unknown by placing 5 cubes in a bag with 7 cubes outside the bag without being observed. Ask, *How can we use balance to figure out the value of the unknown?* (Subtract 7 from both sides.) *What number sentence can we write? How can we check our answer?* (Substitute 5 for the unknown in the equation.)

2 Write $\square + \square + 5 = \square + 13$ on the board. Represent the unknowns by using 3 bags with 8 cubes in each. Explain that the number of cubes in each is the same. Draw an identical symbol, such as $\square$, on each bag to reinforce the idea that the number in each bag is the same. Encourage students to describe how they could use the scale, 3 bags and 18 cubes to help figure out the missing value. Ask them to explain how they know the unknowns represent the same value. Ask, *How do you know?* (The unknowns are identical.) *How can we figure out the unknown value?* Reinforce the idea that the same amount can be subtracted from each side. (Subtract 5 from both sides and subtract the same unknown from both sides.)

3 Repeat the discussion using 3 bags of 12 cubes on one side of the scale and 36 cubes placed on the other side. Ask, *What is known (unknown)? What number sentence can we write?* ($3 \times \square = 36$) *How can we figure out the unknown value?* (Think: "What can I multiply by 3 to get 36?" or "If I split 36 among 3, how many is that for each?") *How can we figure out the unknown in a multiplication sentence? What is the value of the unknown?* (12) *How can we check our answer?*

4 Read the blackline master with the class. Make sure they understand the instructions. Explain that for Questions 3, 5, and 6 they will need to subtract *and* divide to find the unknown value. Allow time for the students to complete the blackline master. Call on volunteers to share their solutions.

Finding Missing Values

Name _____

Subtract or divide cubes or bags to find the unknown value. Use cubes to help you.
Then complete the equation.

1.

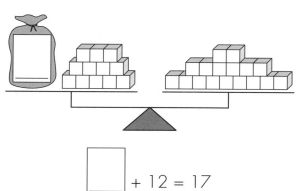

☐ + 12 = 17

2.

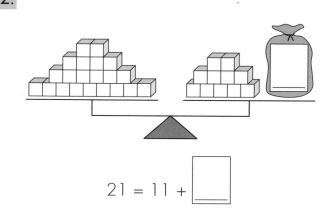

21 = 11 + ☐

3.

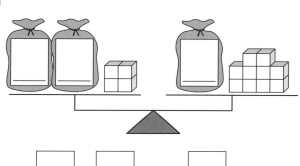

☐ + ☐ + 4 = ☐ + 10

4.

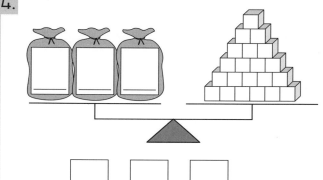

☐ + ☐ + ☐ = 21

5.

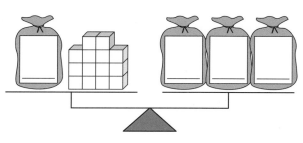

☐ + 14 = ☐ + ☐ + ☐

6.

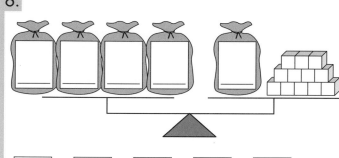

☐ + ☐ + ☐ + ☐ = ☐ + 12

Make a List

Using lists to find two unknowns in simple addition and subtraction equations

AIM

Students will construct a table, make a list, think logically, use "trial and error" or "guess and check" to figure out unknown numbers in a simple system of equations.

MATERIALS

- 1 copy of the blackline master (opposite) for each student

REFLECTION

Refer to the blackline master and encourage confident individuals to explain how they added the same amount to both sides which "removed" an unknown and made it possible to find the value of the unknown that remained. For example "$\bigcirc - \triangle$" and "3" are the same so they can be added to each side of $\triangle + \bigcirc = 35$, as shown below. $\triangle + \bigcirc + \bigcirc - \triangle = 35 + 3$ or $\bigcirc + \bigcirc = 38$. The new equation is still equal.

1 On the board, write the equations shown below.

$$\triangle + \bigcirc = 16 \qquad \bigcirc - \triangle = 2$$

Say, *The total of the values for the triangle and the circle is 16. The value of the circle less the value of the triangle is 2. How can we figure out the value of each unknown?* Ensure the students can logically explain their solutions and that they do not simply guess and check. Encourage them to suggest strategies such as:

- Make a table or list.
  In a table, list all the pairs that add to 16. Identify a pair of numbers that have a difference of 2 such as 7 and 9.

- Work logically.
  The difference is small so the numbers are close together. That means each number is about half of 16.

2 Together read the blackline master. Ask the students to complete the questions. Call on volunteers to share how they figured out the solutions. Discuss how these equations can be solved by substitution, adding the two equations, or systematically completing a table or list.

Make a List

Name _____

For each of these, write values for the shapes to make the equations true. Make a list to help you.

1.

 △ + ◯ = 35   ◯ − △ = 3

┌─────────────────────────┐
│   My List   │
│   │
│   │
│   │
│   │
│   │
└─────────────────────────┘

 △ = _____   ◯ = _____

2.

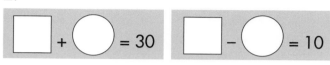 ☐ + ◯ = 30   ☐ − ◯ = 10

┌─────────────────────────┐
│   My List   │
│   │
│   │
│   │
│   │
│   │
└─────────────────────────┘

☐ = _____   ◯ = _____

3.

◯ − △ = 17   ◯ + △ = 59

┌─────────────────────────┐
│   My List   │
│   │
│   │
│   │
│   │
│   │
└─────────────────────────┘

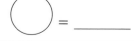 ◯ = _____   △ = _____

4.

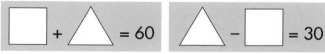 ☐ + △ = 60   △ − ☐ = 30

┌─────────────────────────┐
│   My List   │
│   │
│   │
│   │
│   │
│   │
└─────────────────────────┘

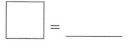 ☐ = _____   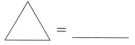 △ = _____

Perfect Punch

Using repeating patterns and tables to examine proportion with two components

AIM

Students will examine real-world proportion problems with two components and represent them as repeating patterns. They will also use these representations to solve more complex problems.

MATERIALS

- 1 copy of the blackline master (opposite) for each student

REFLECTION

Refer to the blackline master and ask, *What do we compare when we are figuring out solutions for these questions? What type of patterns are these?*

1 Read Question 1 on the blackline master with the class. Ask, *What are the 2 ingredients for this punch?* (Orange juice and cranberry juice.) *Do we use the same amounts of each juice?* (No.) *How much of each do we use to make one bowl of punch? If we want to make twice as much punch, how many cups of orange juice do we need? How many cups of cranberry juice do we need? Let's use letters to model this pattern.* Using the starting letter of each juice, write the pattern **OOOOOC OOOOOC** on the board. Ask, *What type of pattern is this?* (A repeating pattern.)

2 On the board, draw the table from Question 1 and complete the first 2 rows with the class, as shown below.

Number of bowls (repeats)	Cups of orange juice	Cups of cranberry juice
1	5	1
2	10	2

Direct the students to complete the table. Ask, *How does the number of cups of orange juice change with the number of bowls? How does the number of cups of cranberry juice change with the number of bowls? What patterns can you see across the table?*

3 Read Question 2 with the class. Then ask the students to complete the question. Call on volunteers to share their answers. Lead a discussion with questions such as, *How does the number of cups of raspberry juice change with the number of bowls? How does the number of cups of grape juice change with the number of bowls? If the number of cups of grape juice is 36, how many bowls of punch will there be? How many cups of raspberry juice will there be?*

Perfect Punch

Name _____

1. To make one bowl of tropical punch, mix 5 cups of orange juice with 1 cup of cranberry juice.
   Complete this table.

Number of bowls (repeats)	Cups of orange juice	Cups of cranberry juice
1		
2		
3		
4		
5		
10		
		12

2. To make one bowl of berry punch, mix 4 cups of raspberry juice with 2 cups of grape juice.

   a. Use **R** for raspberry and **G** for grape. Write 3 parts in the repeating pattern for this punch.

   _____   _____   _____

   b. Complete this table.

Number of bowls (repeats)	Cups of raspberry juice	Cups of grape juice
1		
2		
3		
4		
5		
10		
		32

In the Kitchen

Using repeating patterns and tables to examine proportion with three components

AIM

Students will examine real-world proportion problems with three components and represent them as repeating patterns. They will also use these representations to solve more complex problems.

MATERIALS

- 1 copy of the blackline master (opposite) for each student

REFLECTION

Refer to the blackline master and ask, *What do we compare when we are figuring out solutions for these questions? What type of patterns are these?*

1 Read Question 1 on the blackline master with the class. Ask, *What are the 3 ingredients for these muffins?* (Sugar, butter and flour.) *Do we use the same amounts of each ingredient?* (No.) *How much of each ingredient do we use to make one batch of muffins? If we make 2 batches, how many cups of sugar will we need? How many cups of butter will we need? How many cups of flour will we need? Let's use letters to model this pattern.* Using the first letter of each ingredient, write the pattern **SBBFFFF SBBFFFF** on the board. Ask, *What type of pattern is this?* (A repeating pattern.)

2 On the board, draw the table from Question 1 and complete the first 2 rows with the class, as shown below.

Number of batches (repeats)	Cups of sugar	Cups of butter	Cups of flour
1	1	2	4
2	2	4	8

Direct the students to complete the blackline master. Ask, *How does the number of cups of sugar change with the number of batches? How does the number of cups of butter change with the number of cups of flour? What patterns can you see across the table?*

3 Read Question 2 with the class. Then ask the students to complete the question. Call on volunteers to share their answers. Lead a discussion with questions such as, *How does the number of cups of flour change with the number of batches? How does the number of cups of milk change with the number of tablespoons of butter? If the number of cups of flour is 36, how many batches will there be? How many cups of milk will there be?*

In the Kitchen

Name _____

1. One batch of muffins needs 1 cup of sugar, 2 cups of butter, and 4 cups of flour. Complete this table.

Number of batches (repeats)	Cups of sugar	Cups of butter	Cups of flour
1			
2			
3			
4			
5			
10			
		32	

2. One batch of scones needs 3 cups of flour, 2 tablespoons of butter, and 1 cup of milk.

   a. Use the first letter of each ingredient and write 3 parts in the repeating pattern.

   _____   _____   _____

   b. Complete the table below.

Number of batches (repeats)	Cups of flour	Tablespoons of butter	Cups of milk
1			
2			
3			
4			
5			
10			
		30	

This and That

Considering the multiplicative relationship between two attributes
as repeating patterns

AIM

Students will consider the relationship between two different attributes of repeating patterns as functions.

MATERIALS

- 1 copy of the blackline master (opposite) for each student

REFLECTION

Refer to Question 2 on the blackline master and ask, *How did you use the number of legs to figure out the number of eyes?* (Divided by 2.) *How did you use the number of eyes to figure out the number of legs?* (Multiplied by 2.) Discuss how multiplication and division are inverse operations.

1 On the board, draw and label the table shown below.

Number of hands	1	2	3	4
Number of fingers				

Ask volunteers to stand one at a time and hold up one hand. Record the number of fingers. After four columns have been completed, ask, *How many fingers will we have with 5 hands (6 hands, 10 hands, 20 hands)? How do you know?* (Multiply the number of hands by 5.) Encourage students to describe how they multiplied.

2 Read Question 1 on the blackline master with the class. On the board, draw a kitchen table with 4 chairs. Ask, *If there are 2 separate tables, how many chairs will there be? If there are 4 separate tables, how many chairs will there be? If there are 25 separate tables, how many chairs will there be?* (100) *If there are 120 chairs, how many separate tables will there be?* (30) *How did you figure it out?* Ask the students to complete the question. Call on volunteers to share and justify their answers.

3 Read Question 2 with the class. Have students work independently to complete the table. They can draw pictures of dogs on a separate piece of paper as required. Discuss the thinking they used to figure out the answers. Then ask, *If there are 80 legs, how many dogs will there be?* (20) *How many eyes will there be?* (40) Complete the question as a class. Call on volunteers to share their answers and the thinking they used.

This and That

1. There are 4 chairs around each table.

   a. Complete this chart.

Number of tables	1	2	3	4	5	6
Number of chairs						

   b. If there are 30 tables, there will be _____ chairs.

   Write how you figured it out. _____

   _____

   c. If there are 160 chairs, there will be _____ tables.

   Write how you figured it out. _____

   _____

2. At the pet store, there are lots of dogs.

   a. Complete this chart.

Number of legs	4	8	12	16	20	24
Number of eyes						

   b. If there are 40 legs, there will be _____ eyes, and _____ dogs.

   Write how you figured it out. _____

   _____

   c. If there are 50 eyes, there will be _____ legs, and _____ dogs.

   Write how you figured it out. _____

   _____

Building Patterns

Considering the multiplicative relationship between two attributes
as growing patterns

AIM

Students will consider the relationship
between two different attributes of
growing patterns as functions.

MATERIALS

- Connecting cubes for each
  student and for demonstration

- 1 copy of the blackline master
  (opposite) for each student

REFLECTION

Refer to the blackline master
and invite individuals to describe
the patterns in the charts and
the strategies they used to
figure out the number of chairs
and tables to complete the
statements in each question.

1 On the board, draw the chart and the first picture of a table and chairs
shown below.

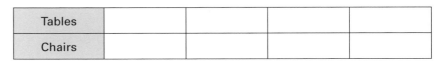

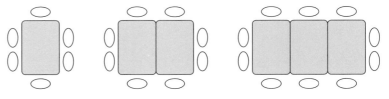

Work with the class to write the numbers in the chart and then draw the
second picture of tables and chairs. Ask, *What should we write in the
chart for this picture? How many tables and chairs do you think will be in
the next picture?* Invite a volunteer to draw the picture, as shown above.

2 Repeat the discussion for 2 more pictures. Then ask, *How many chairs
do you think we will need for 10 tables?* (24) *How do you know?*
Encourage the students to use the pattern and describe a rule for finding
the number of chairs, such as "double the number of tables and add 4".

3 Have the students complete the blackline master.

Building Patterns

Name _____

1. Each side of a table has 1 chair.
   Two or more tables are joined together.

 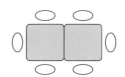 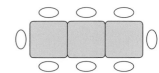

a. Complete this chart.

Number of tables	1	2	3	4	5	6
Number of chairs						

b. If there are 10 tables, there will be _____ chairs.

   Write how you figured it out. _____

   _____

c. If there are 42 chairs, there will be _____ tables.

   Write how you figured it out. _____

   _____

2. Each side of these
   tables has 2 chairs.

 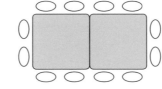 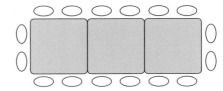

a. Complete this chart.

Number of tables	4	8	12	16	20	24
Number of chairs						

b. If there are 10 tables, there will be _____ chairs.

   Write how you figured it out. _____

   _____

c. If there are 64 chairs, there will be _____ tables.

   Write how you figured it out. _____

   _____

In and Out

Exploring backtracking for simple multiplication functions

AIM

Students will explore backtracking to solve functions when given the output number and a single multiplication rule.

MATERIALS

- 1 copy of the blackline master (opposite) for each student

REFLECTION

Ask, *If we know the IN number and the rule is to multiply by 10, how do we figure out the OUT number?* (Multiply by 10.) *If we know the OUT number, how do we backtrack to figure out the IN number?* (Divide by 10.) *Why?* (Because division is the inverse of multiplication.)

1 On the board, draw a picture of an apple with a $3 price tag and a pineapple with an $8 price tag. Ask, *If we buy apples, how do we figure out the total cost?* (Multiply the number of apples by 3.) On the board, draw the diagram shown below.

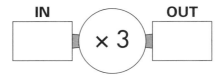

Ask, *If we buy 6 apples, how much will we spend? What is the IN number?* (6) *What is the OUT number?* (18) Write the answers in the diagram then ask, *How can we write this as an equation?* (6 × 3 = 18) Write the equation below the diagram. Repeat for 7 pineapples.

2 Ask, *If we spent some money on apples, how do we figure out how many apples we bought?* (Backtrack and divide the money spent by the price of 1 apple.) *If we spent $96 on some apples, how many did we buy? How did you figure it out?* (Backtracked and divided by 3.) *If we spent $138 on some apples, how many did we buy?* (46) *How do you know?* (138 ÷ 3 = 46) Encourage the students to explain their thinking. Repeat for pineapples, spending $96 and then $136.

3 Have the students complete the blackline master. Allow time for them to share and explain their answers.

In and Out

Name _____

Read the rule. Then write the missing IN and OUT numbers.

1.

IN		OUT
6	×5	
		25
		45
7		
2		

2.

IN		OUT
	×7	21
4		
		63
		49
5		

3.

IN		OUT
8	×3	
		15
		27
11		
		18

4.

IN		OUT
5	×9	
		81
6		
3		
		99

5.

IN		OUT
	×6	36
		54
5		
		18
7		

6.

IN		OUT
8	×2	
		30
		48
32		
		84

7.

IN		OUT
	×8	56
3		
		64
5		
11		

8.

IN		OUT
7	×4	
		36
15		
0		
		48

9.

IN		OUT
	×11	44
		99
3		
10		
		121

Which Way?

Exploring different ways of solving simple multiplication problems

AIM

Students will explore some different ways of solving simple multiplication problems. They will also review the inverse relationship between multiplication and division.

MATERIALS

- 1 copy of the blackline master (opposite) for each student

REFLECTION

Ask, *What is another real-world problem that is a function with a multiplication rule?* If necessary, suggest, *Imagine we are buying a number of stamps and the price of 1 stamp is 50 cents.* Discuss the different ways to represent this problem: a table of values, function machine, arrow mathematics, or a rule.

---

1 Ask, *If 1 CD costs $5, how can we figure out the cost of 9 CDs?* Elicit several suggestions and then discuss and work through each method with the class.

- On the board, draw the table shown below and work with the students to complete the table.

Number of CDs	1	2	3	4	5	6	7	8	9
Total cost									

- Draw a function machine on the board, as shown. Ask, *What is the rule for this function machine?* (Multiply by 5.) *If 9 is the IN number, what will be the OUT number?* (45)

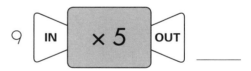

- Write a rule that matches the problem. Ask, *If we buy CDs, how do we figure out the total cost? How can we write this as a rule?* (Number of CDs × $5 = Total cost.)

2 Refer to the function machine and say, *Imagine you spend $35 to buy CDs. Where should we write this number on the function machine? How many CDs did you buy? How do you know?* Invite individuals to describe their thinking.

3 Ask the students to complete the blackline master. Call on volunteers to share their answers.

Which Way?

Name _____

1. A bottling machine fills 6 bottles every minute.

    a. Complete this table to show the number of bottles it fills each minute for up to 9 minutes.

Minutes	1	2	3	4	5	6	7	8	9
Bottles									

    b. Write the rule for figuring out the number of bottles filled if you know the number of minutes.

    _____

    c. Write the rule on this function machine.

    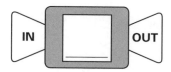

    d. Use your rule to figure out the number of bottles the machine fills in:

    12 minutes = _____ bottles

    40 minutes = _____ bottles

    1 hour = _____ bottles

2. Write the IN number for each of these machines.

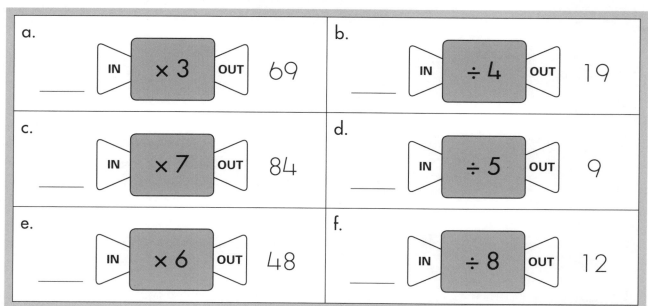

a.

_____ IN ×3 OUT 69

b.

_____ IN ÷4 OUT 19

c.

_____ IN ×7 OUT 84

d.

_____ IN ÷5 OUT 9

e.

_____ IN ×6 OUT 48

f.

_____ IN ÷8 OUT 12

Party Plans

Combining multiplication and addition

AIM

Students will use multiplication and addition to create rules that generate output numbers for given input numbers. They will also backtrack these rules.

MATERIALS

- 1 copy of the blackline master (opposite) for each student

REFLECTION

Discuss rules that the students used to figure out the total for any number of items for each of the examples in Question 1 of the blackline master. Refer to Question 2 and discuss how to check that all possible solutions have been found by constructing an ordered list.

1 On the board, draw the table and illustration shown below.

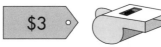

Number of whistles	Total cost

Ask, *If we buy 1 (2, 3, 4) whistle, how much will we spend? How can we use the table to show the total cost for each purchase? If we buy 9 whistles, how much will we spend? ($27) How do you know? What rule can we use to figure out the total cost?*

2 Repeat the discussion for buying hats at $5 each. Then draw the table shown below and repeat the discussion for buying whistles and hats.

Number of whistles	Number of hats	Total cost

Have students work with a partner to find all of the different combinations that total less than $30. Encourage them to suggest a rule and write it in words and symbols, for example, "whistles × 3 plus hats × 5 equals the total cost".

3 Ask the students to complete the blackline master. Call on volunteers to share their solutions.

Party Plans

Name _____

1. Complete the tables below.

1 Bag of balloons = $3	
Bags of balloons	Cost
1	
2	
3	
4	
5	

1 Pizza = $7	
Pizzas	Cost
1	
2	
3	
4	
5	

1 Box of cola = $8	
Boxes of cola	Cost
1	
2	
3	
4	
5	

Patterns and Functions

7

2. Josh is shopping for a party. He spends less than $30 and buys some balloons, pizzas, and cola. Use the table below to figure out what he may have bought.

Bags of balloons	Cost	Pizzas	Cost	Boxes of Cola	Cost	Total Cost
1	$3	1	$7	1	$8	$18

Follow the Pattern

Expressing growing patterns as equations

AIM

Students will use equations to write rules to describe growing patterns using two operations.

MATERIALS

- Square tiles
- 1 copy of the blackline master (opposite) for each student

REFLECTION

Refer to each of the patterns in Question 1 on the blackline master and ask the students to describe how the pictures are changing and how they figured out the numbers to complete each sentence. Invite volunteers to explain how they decided what to draw and write for each part of Question 2.

1 Use tiles to make the pattern shown below or draw and label the pictures on the board.

Picture 1   *Picture 2*   *Picture 3*   *Picture 4*

Ask, *What do you notice about this pattern? What is changing? What stays the same? How many tiles is the pattern growing by? What will the next picture look like?*

2 Point to Picture 1 and ask, *What do you notice about this picture? How many columns (rows) do you see? How many tiles are there in total? How can we write this as a number sentence?* (2 + 1 = 3) Point to Picture 2 and ask, *What stayed the same as the first picture?* (The number of rows.) *What changed?* (The number of columns and the total number.) *How did they change? What equation can we write?* Invite volunteers to describe their thinking, then write **2 + 2 + 1 = 5** and **2 x 2 + 1 = 5** on the board. Ask the students to suggest equations for the next 2 pictures. (2 × 3 + 1 = 7 or 3 × 2 + 1 = 7 *and* 2 × 4 + 1 = 9 or 4 × 2 + 1 = 9.) Encourage them to identify the picture number in each equation. Call on a volunteer to make the next part. Ask, *What equation can we write for this picture?* (2 × 5 + 1 = 11) *How do you know?* Discuss how a pattern rule helps you figure out how many tiles will be in any part of the pattern. Ask, *What rule can we write for this pattern?* (2 × picture number + 1 = total *or* picture number × 2 + 1 = total.)

3 Ask the students to complete the blackline master. Call on volunteers to share their answers. Discuss the patterns in the equations.

Follow the Pattern

Name _____

1. Draw the next picture for each pattern. Complete the number sentences to match.

Pattern A	Pattern B
Picture 1	Picture 1
$(1 \times 3) + 1 = $ _____	$1 + (1 \times 2) = $ _____
Picture 2	Picture 2 
$(2 \times 3) + 1 = $ _____	$1 + (2 \times 2) = $ _____
Picture 3	Picture 3 
$(3 \times $ _____ $) + $ _____ $ = $ _____	$1 + ($ _____ $ \times 2) = $ _____
Picture 4	Picture 4
$($ _____ $ \times $ _____ $) + $ _____ $ = $ _____	_____ $ + ($ _____ $ \times $ _____ $) = $ _____

2. Draw the 8th picture for each pattern. Write a matching number sentence.

Picture 8	Picture 8
$($ _____ $ \times $ _____ $) + $ _____ $ = $ _____	_____ $ + ($ _____ $ \times $ _____ $) = $ _____

For Sale

Using function machines to model two-step real-world problems

AIM

Students will use 2 operations, and a division rule and then a multiplication rule to calculate output numbers when given input numbers.

MATERIALS

- 1 copy of the blackline master (opposite) for each student
- Calculator for each student

REFLECTION

Refer to Question 3 on the blackline master and ask, *What operation did you use to figure out the number of packs?* (Division.) *What operation did you use to figure out the total dollar value?* (Multiplication.) *If the storekeeper sold some muffins for $60, how many packs did she sell?* (50) *How many muffins did she sell?* (150) *How did you figure out the answers?* (Used inverse operations: division then multiplication.)

1 On the board, draw the function machine picture shown below, without any labels.

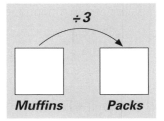

Refer to Question 1 on the blackline master and ask, *If there are 30 (120, 96) muffins, how many packs will there be?* (10, 40, 32) Call on volunteers to share the thinking they used. Then ask, *What can we write on this In/Out machine to show our thinking?* Work with the class to write the words and rule shown.

2 Erase the labels on the function machine and repeat the discussion for cookies and cakes using mentally manageable multiples of 6 and 5 respectively. Ask the students to work independently to complete Question 1.

3 Discuss the first example in Question 2. Say, *Cookies cost $1.50 for a pack of 6. How many packs will you need to get 48 cookies?* (8) *What is the total cost of these packs?* (8 packs cost $12.) *How do you know?* Invite volunteers to describe the thinking they used. During the discussion, extend the function machine above to show the picture below.

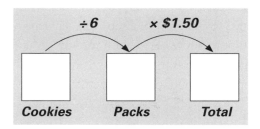

4 Ask the students to work independently to complete Question 3. Some students may need to use a calculator for some parts of this question.

For Sale

Name _____

Muffins are sold in packs of 3, cookies are sold in packs of 6, and cakes are sold in packs of 5.

1. Complete the tables below.

Muffins	Packs of 3
12	4
45	
66	
123	
150	

Cookies	Packs of 6
12	2
36	
126	
108	
312	

Cakes	Packs of 5
10	2
85	
125	
90	
200	

2. Cookies cost $1.50 per pack. Complete this table to show the total dollar value for each number of cookies.

Cookies	Packs	Cost per pack	Total dollar value
36	6	$1.50	$9.00
72		$1.50	
108		$1.50	
120		$1.50	
318		$1.50	

3. Muffins cost $1.20 per pack. Complete this table to show the total dollar value for each number of muffins.

Muffins	Packs of 3	Cost per pack	Total dollar value
36		$1.20	
81		$1.20	
90		$1.20	
120		$1.20	
180		$1.20	

Buying and Selling

Reinforcing the relationship between multiplication and division

AIM

Students will use multiplication and division to calculate output/input numbers for given input/output numbers. This activity will reinforce the idea that multiplication and division are inverse operations.

MATERIALS

- 1 copy of the blackline master (opposite) for each student

REFLECTION

Refer to Question 3 on the blackline master and ask, *What is the inverse of multiplication and then division?* (Multiplication and then division.) *What is the inverse of division and then multiplication?* (Division and then multiplication.)

1 Draw the first function machine from the previous activity (For Sale) on the board. Say, *Muffins are sold in packs of 4. If we buy 7 packs, how many muffins will we have?* (28) *How do you know?* Write **7** in the Out box and encourage the students to explain how they would use multiplication. Repeat the discussion for other numbers of packs and draw an arrow below the boxes. Write the rule as shown below.

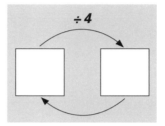

2 On the board, draw the function machine shown below, without the backtrack arrows underneath.

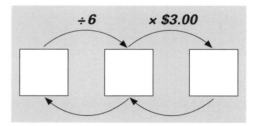

Explain that rolls are sold in packs of 6 and cost $3 a pack. Ask, *If we need 24 rolls, how many packs should we buy and how much will they cost in total? How did you figure it out?* (Divide by 6 and multiply by 3.) *If we spend $30 on rolls, how many will we buy? How did you figure it out?* (Divide by 3 and then multiply by 6.) Encourage volunteers to explain how they used the related operations and draw the backtrack arrows below the boxes.

3 Have the students complete the blackline master. Allow time for them to share and explain their answers. Discuss how multiplication and then division is the inverse of multiplication and then division.

Buying and Selling

Name _____

1. Burgers cost $6 each, pizzas cost $7 each, and salads cost $8 each.
    Complete the tables below.

Burgers	Cost
4	
	$84
13	
	$60
8	

Pizzas	Cost
8	
	$77
15	
	$63
7	

Salads	Cost
7	
	$160
13	
	$112
5	

2. Muffins are sold in packs of 3, cookies are sold in packs of 6, and cakes are sold in packs of 2.
    Complete the tables below.

Muffins	Packs of 3
12	4
	12
48	
	23
123	

Cookies	Packs of 6
24	
	12
96	
	14
48	

Cakes	Packs of 2
122	
	48
56	
	120
36	

3. Cookies are sold in packs of 6 and cost $1.50 per pack. Complete the table below.

Cookies	Packs of 6	Cost per pack	Total dollar value
24		$1.50	
	10	$1.50	
		$1.50	$10.50
54		$1.50	
	8	$1.50	

Go Grids

Using array models to reinforce turnarounds for multiplication

AIM

Students will explore array models to establish that the same answer results when the order of multiplication is changed.

MATERIALS

- Grid paper for each student
- Scissors for each student
- 1 copy of the blackline master (opposite) for each student
- Calculator for each student

REFLECTION

Ask, *When we multiply 2 numbers, does it matter which order we use to multiply them? Will this be true for large numbers?* Ask the students to suggest pairs of large numbers. List the numbers on the board. Have them use a calculator to multiply each pair of numbers in 2 different orders. Ask, *Can you think of another operation where we can change the order and the answer will remain the same?* (Addition.)

1 Using grid paper, cut out the grid shown below.

Ask, *What is the number of squares (area) in this grid? How can we figure it out?* (Multiply.) Rotate the grid a one-quarter turn and ask, *When we rotate the grid a one-quarter turn, what do you notice? What is the same (different)?* Reinforce the idea that the number of squares has not changed, but the array can be described another way. (7 rows of 4.) Ask, *What number sentence can we write for each array?* Write the 2 equations on the board.

2 Provide students with grid paper and scissors. Encourage them to each cut out one array. Each dimension should be between 4 and 9 rows. Invite volunteers to show their grids and describe the arrangements and the equations they can write for each, for example, $8 \times 9 = 72$ and $9 \times 8 = 72$. Reinforce the use of "turnaround" to describe the related number sentences. This is a good opportunity to challenge the students to find arrays that look the same when they are rotated.

3 Ask the students to complete the blackline master. Call on volunteers to share their answers.

Go Grids

Name _____

1. Write 2 turnaround multiplication facts that can be used to figure out the area of each grid.

a. 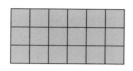 3 × _____ = _____ _____ × 3 = _____	**b.** 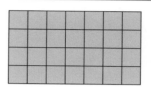 4 × _____ = _____ _____ × _____ = _____
c. 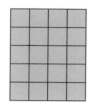 5 × _____ = _____ _____ × _____ = _____	**d.**  _____ × _____ = _____ _____ × _____ = _____

2. Draw grids to show each expression. Then write 2 turnaround multiplication facts to match.

a. 7 × 3 _____ × _____ = _____ _____ × _____ = _____	**b.** 9 × 2 _____ × _____ = _____ _____ × _____ = _____
c. 6 × 5 _____ × _____ = _____ _____ × _____ = _____	**d.** 4 × 6 _____ × _____ = _____ _____ × _____ = _____

Properties

1

Does it Commute?

Exploring the commutative property for addition and subtraction involving decimal fractions

AIM

Students will use a number line to explore the commutative property for addition and subtraction. They will discover the commutative property is not true for subtraction.

MATERIALS

- 1 copy of the blackline master (opposite) for each student

REFLECTION

Discuss how the answer remains the same if 2 decimal numbers are added in any order but if 2 decimal numbers are subtracted in any order, the answers are different.

1   Ask, *If you run, then skip 1.7 metres and then jump 2.4 metres, what is the total length of your "skip and jump"? How can we figure out the total length? Does it matter which order we use to add the two lengths?*

2   On the board, write **1.7 + 2.4** and draw the number line shown below.

Ask, *How can we add the numbers on the number line?* Elicit several responses and then invite a student to draw the jumps on the number line. Ask, *Will it matter if we add 1.7 (2.4) first?* Invite a second student to draw jumps to model the turnaround. Ask, *Does it matter which number we add first? Is the answer the same for both orders? How can we write this?* Write **2.4 + 1.7 = 1.7 + 2.4** on the board. Ask, *What do we call this?* (A turnaround.) Repeat for other pairs of decimals that the students suggest.

3   On the board, write **7.2 – 3.8** and draw the number line shown below.

Ask, *How do we show 7.2 subtract 3.8 on the number line?* Invite a student to draw jumps on the number line to show 7.2 – 3.8. Ask, *What happens if we start at 3.8 and subtract 7.2? Does it matter if we start with 7.2 or 3.8?* Invite a second second student to draw jumps to model the turnaround 3.8 – 7.2 to show that the answers are not the same. If time allows, repeat the discussion for 8.6 – 5.7.

4   Ask the students to complete the blackline master. Call on volunteers to share their answers.

Does it Commute?

Name _____

1. Write the answer to each equation and its turnaround. Draw jumps to help you.

a.

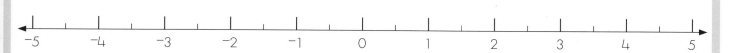

3.5 + 1.5 = _____   1.5 + 3.5 = _____

b.

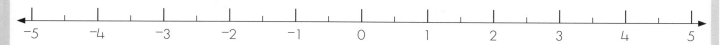

3.5 − 1.5 = _____   1.5 − 3.5 = _____

c.

2 + 2.5 = _____   2.5 + 2 = _____

d.

2 − 2.5 = _____   2.5 − 2 = _____

2. Place a ✓ on each pair of equations above that have the same answer. Write what you notice.

_____

_____

Fair Share

Exploring the commutative property for division

AIM

Students will use real-world examples to discover that the commutative property is not true for division.

MATERIALS

- 1 copy of the blackline master (opposite) for each student

REFLECTION

Write **12 ÷ 4 = ___** and **4 ÷ 12 = ___** on the board and ask, *What do you know about the answers to these number sentences? Are the answers equal? Which answer is greater? How do you know?*

1 Draw 20 apples on the board and say, *We have 20 apples to share among 5 people. How many apples will we give each person? How did you figure it out?* Model the sharing with the apples on the board. Ask, *How do we write this?* Write **20 ÷ 5 = 4** on the board. Ask, *If we have 5 apples to share among 20 people, how many apples will we give each person? What do we know about the answer? Will we give them more than 1 or less than 1 each? How do you know? How can we write 5 shared among 20?* On the board, draw 5 apples. Invite confident individuals to describe how they could share the apples by drawing lines to cut them into quarters. Then write $5 ÷ 20 = \frac{1}{4}$ on the board. Ask, *What number sentence can we write about the 2 equations?* Invite several responses then write **20 ÷ 5 ≠ 5 ÷ 20** on the board. Repeat for 15 oranges shared among 5 people, and 5 oranges shared among 15 people.

2 Have the students complete the blackline master. Call on volunteers to share their answers.

Fair Share

1. **a.** Draw lines to share 8 apples among 4 people. Complete the equation.

_____ ÷ _____ = _____

b. Draw lines to share 4 apples among 8 people. Complete the equation.

_____ ÷ _____ = _____

c. Write **=** or **≠** to complete this number sentence.   8 ÷ 4 _____ 4 ÷ 8

2. **a.** Draw lines to share 9 pears among 3 people. Complete the equation.

_____ ÷ _____ = _____

b. Draw lines to share 3 pears among 9 people. Complete the equation.

_____ ÷ _____ = _____

c. Write **=** or **≠** to complete this number sentence.   9 ÷ 3 _____ 3 ÷ 9

3. Write **=** or **≠** to complete each of these.

a. 12 ÷ 4 _____ 4 ÷ 12

b. 5 ÷ 10 _____ 10 ÷ 5

c. 4 ÷ 16 _____ 16 ÷ 4

d. 18 ÷ 6 _____ 6 ÷ 18

Making Money

Establishing that zero is the identity for addition and subtraction

AIM

Students will interpret real-world information and establish that addition or subtraction involving zero does not change the "starting number". The term "identity" is used to describe zero in an addition or subtraction sentence.

MATERIALS

- 1 copy of the blackline master (opposite) for each student
- Calculator for each student

REFLECTION

Ask, *When we add or subtract zero, what happens to the starting number?* (It remains the same or is identical.) Introduce the term "identity" and say, *Zero is the identity for addition and subtraction. Why do you think the term "identity" is used?* (Because adding or subtracting zero leaves the starting amount identical.)

1 Draw a piggy bank on the board and say, *Imagine we have some money saved in our piggy bank. If we don't save any more money this week, how much money will we have? How can we write this?* Write ___ *+ 0 =* ___ on the board. Ask, *If we don't spend any money from our savings this week, how much money will we have? How can we write this?* Write ___ *– 0 =* ___ on the board. Ask, *When we add or subtract zero, what happens to the starting number?* (It remains the same.)

2 On the board, draw the table from the blackline master. Refer to the timeline and ask, *At 8 a.m., how much money does the store have? Did the store sell any items at 8 a.m.? How do we record this in the table?* Record **$235** and the equation **235 + 0 = 235** in the table. Ask, *What happens at 9 a.m.?* (1 office chair is sold.) *How much money does the store have now?* ($320) Record **$320** and write the equation, as shown below.

Time	Equation	Total $
8 a.m.	235 + 0 = 235	$235
9 a.m.	235 + 85 = 320	$320

3 Direct the students to complete the table on the blackline master. Students may use their calculators if necessary. Allow time for them to share and discuss their answers.

Making Money

Name _____

$215

$85

$12

$176

$321

A storekeeper opens with $235 cash. At midday each day, the storekeeper buys lunch for $12. The timeline below shows the sales that were made on Friday.

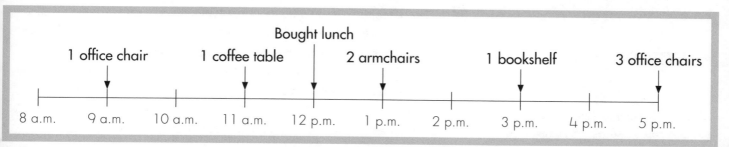

Complete the table below to show how much money the store has at each working hour on Friday.

Time	Equation	Total $
8 a.m.	235 + 0 = 235	$235
9 a.m.		
10 a.m.		
11 a.m.		
12 p.m.		
1 p.m.		
2 p.m.		
3 p.m.		
4 p.m.		
5 p.m.		

Properties

4

Buying and Sharing

Exploring the order of operations involving multiplication and division

AIM

Students will use appropriate whole numbers to complete number sentences involving multiplication and division.

MATERIALS

- 1 copy of the blackline master (opposite) for each student

REFLECTION

Discuss the steps for completing number sentences that involve multiplication and division. Reinforce the idea that these can be completed in any order.

1 On the board, draw a bunch of 6 flowers. Say, *The flower shop sells flowers in bunches of 6. Chloe buys 10 bunches. How many flowers does she have? How can we write this?* On the board, write **Chloe's flowers = 6 × 10**. Ask, *If Chloe shares the flowers equally with a friend, how many flowers will Chloe have in her share? How can we write this?* Write **Chloe's share of flowers = 6 × 10 ÷ 2** on the board. Invite students to describe how they can figure out the number of flowers in each share using multiplication and then division or division and then multiplication.

2 Say, *Kieran is given $5 every week to save. How much money will he have after 9 weeks?* Encourage responses and on the board, write **Kieran's savings = 5 × 9**. Ask, *If Kieran uses all this money to buy 3 tickets, how much does each ticket cost? How can we write this?* Write **Price of tickets = 5 × 9 ÷ 3** on the board. Invite students to describe how they can figure out the price for each ticket using multiplication and then division or division and then multiplication.

3 Ask the students to complete the blackline master. Call on volunteers to share their answers.

Buying and Sharing

Name _____

1. Haley buys 9 bunches of 8 flowers.

a.	Complete a matching equation.
	Haley's flowers = _____ × _____
b.	Haley shares her flowers equally with her 3 friends. Complete the equation to show Haley's share of flowers.
	Haley's share = _____ × _____ ÷ _____
c.	Write Haley's share and how you figured it out.
	_____
	_____
	_____

2. Anita saves $15 each week for 4 weeks.

a.	Complete a matching equation.
	Anita's savings = _____ × _____
b.	Anita shares her money equally with her 2 sisters. Complete the equation to show Anita's share of money.
	Anita's share = _____ × _____ ÷ _____
c.	Write Anita's share and how you figured it out.
	_____
	_____
	_____

Properties

5

Carrying Crates

Using different representations to show relationships

AIM

Students will use tables and graphs to represent the relationship between variables. They will also write rules for the relationships.

MATERIALS

- 1 copy of the blackline master (opposite) for each student

REFLECTION

Refer to the blackline master and ask, *If there are 15 vans, how can we figure out how many crates will fit?* Discuss the 3 different ways:

- extend the graph to 15 vans;
- extend the table to 15 vans; or
- substitute 15 in the equation, number of crates = 3 × number of vans.

1 Say, *Each delivery van holds 4 crates. How many crates will fit in 8 vans? How can we figure out the answer?* Elicit responses and then suggest using a table of values to show the numbers of vans and crates. On the board, draw the table shown below, and call on volunteers to complete the values.

Number of vans	1	2	3	4	5	6	7	8	9
Number of crates									

2 Ask, *How can we draw this data as a graph? What will we call the graph? What will we call the two axes?* (Horizontal axis — number of vans, vertical axis — number of crates.) On the board, draw and label the graph. Mark the increments on the horizontal axis from 1 to 10 and on the vertical axis from 1 to 40. Call on volunteers to plot the points from the table.

3 Ask, *What rule can we write for this graph?* (Number of crates = 4 × number of vans.) Encourage a volunteer to write the rule on the board. Ask, *How many crates will fit in 12 vans? How did you figure it out?* Ensure the students explain their thinking.

4 Have the students complete the blackline master. Call on volunteers to share their answers.

Carrying Crates

Name _____

A delivery van holds 3 crates.

1. Complete the table below to show how many crates will fit in up to 10 vans.

Number of vans	1	2	3	4	5	6	7	8	9	10
Number of crates										

2. Write how you can figure out the number of vans when you know the number of crates.

_____

3. Complete these.

   a. 15 vans will hold _____ crates.

   b. 18 vans will hold _____ crates.

   c. 36 crates will fit in _____ vans.

   d. 69 crates will fit in _____ vans.

4. a. Write the data from the table above
      as ordered pairs.

      (____,____) (____,____) (____,____)

      (____,____) (____,____) (____,____)

      (____,____) (____,____) (____,____)

      (____,____)

   b. Plot the ordered pairs above
      onto the graph at right.

   c. Write a name for the graph.

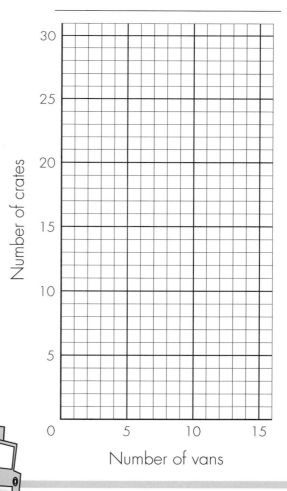

Dotty Data

Working with graphs to represent discrete data

AIM

Students will represent simple multiplication situations as graphs, and make predictions based on the information.

MATERIALS

- 1 copy of the blackline master (opposite) for each student

REFLECTION

Refer to the blackline master and discuss questions such as, *Can you buy an ice-cream cone with $1\frac{1}{2}$ scoops? If you can, where will that be on the graph? Does it make sense to show points between whole numbers of scoops?*

1 Say, *The price of one postage stamp is 50 cents. How can we show the total costs of up to 10 stamps?* Suggest that this data can be represented by a table of values. On the board, draw the table shown below, and ask volunteers to help complete the total costs.

Number of stamps	1	2	3	4	5	6	7	8	9	10
Total cost	$0.50									

2 Draw the graph shown below, on the board.

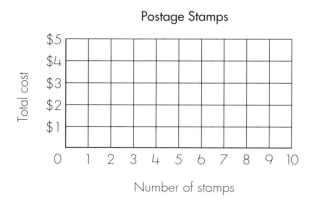

Postage Stamps

Invite volunteers to help plot the points from the table then ask, *Should we join the points?* (No.) *Why not?* (It is not possible to buy $1\frac{1}{2}$ stamps.) *If we know the number of stamps, what rule can we use to figure out the total cost?* (Total cost = 50 × number of stamps.)

3 Extend the x axis to 15 and the y axis to $7.50. Discuss the trend in the graph and ask, *How can we extend the data on this graph? Do we need to figure out the costs? Look at the points on the graph. What pattern can you see?* Use the pattern to extend the data but do not join the points. Ask, *How much do 12 stamps cost?* ($6) *If we spend $7 on 50-cent stamps, how many stamps will we buy?* (14)

4 Have the students complete the blackline master. Call on volunteers to share their answers.

Dotty Data

Name _____

All ice-cream cones have 3 scoops of ice cream.

1. Complete the table below.

Number of cones	1	2	3	4	5	6	7
Number of scoops							

2. Plot these points on this graph.

3. Plot up to 12 ice-cream cones on the graph, and complete these.

   a. 8 ice-cream cones = _____ scoops

   b. 10 ice-cream cones = _____ scoops

   c. 27 scoops = _____ ice-cream cones

   d. 33 scoops = _____ ice-cream cones

   e. 23 ice-cream cones = _____ scoops

   f. 96 scoops = _____ ice-cream cones

4. Write how you can figure out the number of scoops when you know the number of cones.

   _____

   _____

   _____

   _____

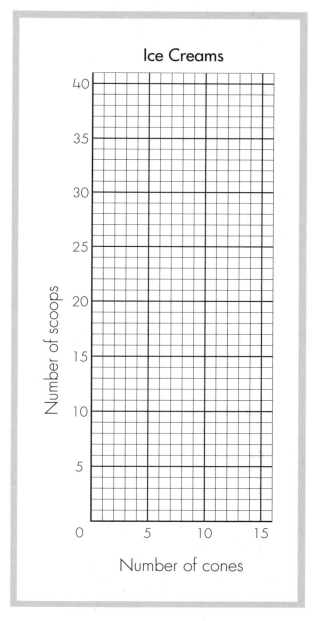

Ice Creams

Representations

2

Dots or Lines

Working with discrete and continuous data

AIM

Students will decide whether points on a graph can be joined.

MATERIALS

- 1 copy of the blackline master (opposite) for each student

REFLECTION

Discuss real-world situations where it would make sense to join the points of a graph, for example, students' heights, the speed of cars. Discuss other real-world situations where it would not make sense, for example, number of pets in each family, number of books in the classroom.

1 On the board, draw a canoe and say, *Canoes are for rent. Each canoe holds 2 people. You are not allowed to rent a canoe if you are by yourself.* Refer to the table on the blackline master, and ask, *If 1 canoe is rented, how many people must there be?* Ask the students to complete the table for up to 5 canoes.

2 Ask, *How can we show this as a graph?* Instruct the students to complete Question 1b. Ask, *If we have a point between 1 and 2 canoes, what does that mean? Can you have $1\frac{1}{2}$ canoes? Does it make sense to join the points?* (No.) *What happens to the number of people as the number of canoes changes? If there are 8 canoes rented, how can we figure out the number of people? How can we write this?* Elicit responses, then write **Number of people = 2 × number of canoes** on the board. Have the students write the equation for Question 1d on the blackline master. Ask, *If we know the number of people, how can we figure out the number of canoes? How can we write this?* Elicit responses, then write **Number of canoes = number of people ÷ 2** on the board.

3 Ask the students to complete Question 2. Call on volunteers to share and justify their answers.

Dots or Lines

Name _____

1. **a.** Each canoe holds 2 people. Complete the table below.

Number of canoes	1	2	3	4	5
Number of people					

b. Plot the points on this graph.

c. Should you join the points? _____

d. Write how you can figure out the number of people when you know the number of canoes.

_____

_____

_____

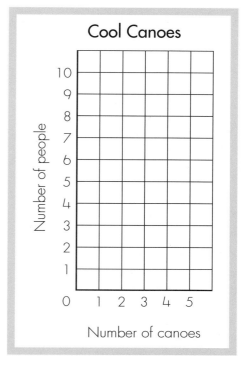

Cool Canoes

(vertical axis: Number of people, 1–10; horizontal axis: Number of canoes, 0 1 2 3 4 5)

2. For each graph, write if you should join the points.

a. There are 5 birds in each tree.

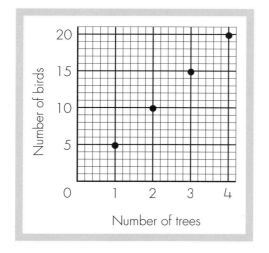

(vertical axis: Number of birds, 5 10 15 20; horizontal axis: Number of trees, 0 1 2 3 4)

Should you join the points? _____

b. Ben saves half of his allowance.

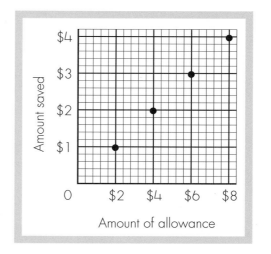

(vertical axis: Amount saved, $1 $2 $3 $4; horizontal axis: Amount of allowance, 0 $2 $4 $6 $8)

Should you join the points? _____

Representations

3

Printing Press

Representing change using continuous straight-line graphs

AIM

Students will be introduced to continuous straight-line graphs as a means of describing change.

MATERIALS

- 1 copy of the blackline master (opposite) for each student

REFLECTION

Ask, *If the 1st press can print 20 pages in 5 seconds, how many pages can it print in 10 seconds? How many pages in 20 seconds? What will this graph look like? Will it be a straight line? Can you join the points? How do you know?*

1 On the board, draw the table shown below.

Time (seconds)	5	10	15	20			
Number of pages							

Read the top of the blackline master with the class. Ask, *How many pages can the press print in 5 seconds?* Direct the students to complete the table on the blackline master. Call on volunteers to share their answers by completing the table on the board.

2 Have the students complete the blackline master. Call on volunteers to share and explain their answers.

3 Draw a 2nd table on the board and say, *Imagine another printing press prints 5 pages in 5 seconds.* Work with the class to complete the table for the 2nd printing press. Have the students plot the points from the 2nd table onto the graph on the blackline master. Say, *Draw another line to join these points. Which line shows the 1st press? Which line shows the 2nd press? Which line is steeper, the line that shows the 1st press or the line that shows the 2nd press?*

Printing Press

Name _____

A printing press prints 10 pages every 5 seconds.

1. Complete the table below.

Time (seconds)	5	10	15	20			
Number of pages							

2. Plot the points on the graph below.

3. Draw a line to join the points.

4. Write how you can figure out the number of pages when you know the number of seconds.

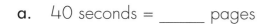

_____

_____

_____

_____

5. Complete these.

   a. 40 seconds = _____ pages

   b. 50 seconds = _____ pages

   c. 30 pages = _____ seconds

   d. 80 pages = _____ seconds

   e. 2 minutes = _____ pages

   f. 3 minutes = _____ pages

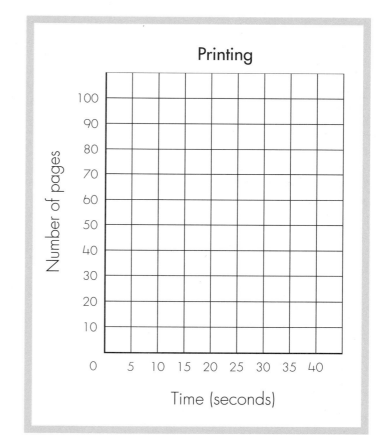

Printing

Representations

4

Rushing Robots

Comparing graphs for different rates

AIM

Students will draw graphs for different rates of change.

MATERIALS

- 1 copy of the blackline master (opposite) for each student

REFLECTION

Refer to the blackline master and discuss questions such as, *What do you notice about the graphs for the 3 robots?* (They are straight lines.) *Which line is steeper (goes up faster)?* (Robot C.)

1 Read the top of the blackline master with the class. Refer to Question 1a and ask, *How many bottles does Robot A pack in 1 minute? How many in 2 minutes? How do you know?* Refer to the table for Robot B and ask, *How many bottles does Robot B pack in 1 minute? How many in 2 minutes? How do you know?* Ask the students to complete Question 1. Call on volunteers to share their answers. Then ask questions such as, *Which robot will be the first to pack 96 bottles? How do you know?* Discuss strategies such as extending the table or the graph, or calculating the number of minutes.

2 Ask the students to complete the blackline master. Discuss the different ways to figure out the graph for Robot C, for example, adding the points for Robot A and Robot B in the vertical direction, or drawing a table of values and plotting the points.

Rushing Robots

Name _____

Robot A packs 6 bottles each minute. Robot B packs 3 bottles each minute.

1. **a.** Complete this table for Robot A.

Number of minutes	1	2	3	4	5	6
Number of bottles						

    b. Complete this table for Robot B.

Number of minutes	1	2	3	4	5	6
Number of bottles						

2. On the graph, plot the points for Robot A in red and the points for Robot B in blue.

3. Complete these.

    a. Robot A packs 48 bottles in _____ minutes.

    b. Robot B packs 27 bottles in _____ minutes.

4. Robot C packs 9 bottles each minute. Will the points for Robot C be **above** or **below** the points on the graph?

    _____

5. Plot 5 points for Robot C in green.

6. Explain how you figured out the points for Robot C.

    _____

    _____

    _____

    _____

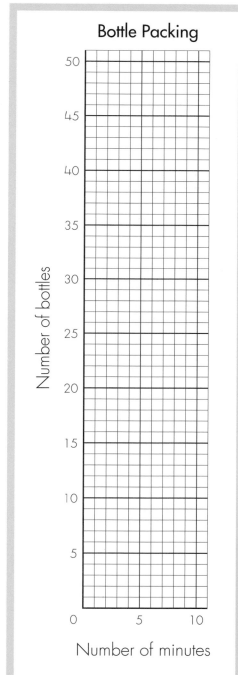

Bottle Packing

Puzzle Graphs

Interpreting points on a scatter plot

AIM

Students will interpret points on a scatter plot and justify their choices.

MATERIALS

- 1 copy of the blackline master (opposite) for each pair of students

REFLECTION

Refer to the blackline master and ask, *How did you figure out what each point represents? Did you focus on a particular point first? Did you focus on a particular animal first? Why did you choose that animal? Was it the fastest (heaviest) animal?*

1 On the board, draw the graph shown below and simple pictures of a truck, a car, a motorcycle, and a bicycle.

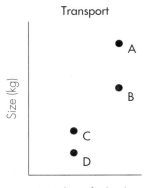

Ask, *What does this graph compare?* (4 different forms of transport.) *What do the points on the graph represent?* (Each is a different vehicle.)

2 Discuss how the points that are at the greatest distance from each axis have the greatest values. Ask, *Which points have the same number of wheels?* (Points A and B, and Points C and D.) *Which points have 2 wheels and which points have 4 wheels?* (Points C and D = 2 wheels, and Points A and B = 4 wheels.) *How do you know? Which vehicle is represented by Point A?* (Truck.) *Which point represents the car?* (Point B.) *Which represents the motorcycle?* (Point C.) *Which vehicle is represented by Point D?* (Bicycle.) *How did you decide which vehicle each point represents?* As the students justify their decisions, list the reasons on the board. They may say, for example, "The truck weighs the most, so it must be Point A. The motorcycle and bicycle both have 2 wheels, and the bicycle is lighter than the motorcycle so it must be Point D. The motorcycle must be Point C. The car has 4 wheels and is lighter than the truck, so it must be Point B".

3 Read the blackline master with the class. Ask the students to work in pairs to complete the page. Call on pairs of volunteers to share and justify their answers.

Puzzle Graphs

This table shows the average speed of 6 animals.
Each animal in the table is represented on the scatter plot below.

Animal	Average speed
Cat	50 km/h
Chicken	17 km/h
Grizzly bear	50 km/h
Hunting dog	62 km/h
Lion	72 km/h
Pig	17 km/h

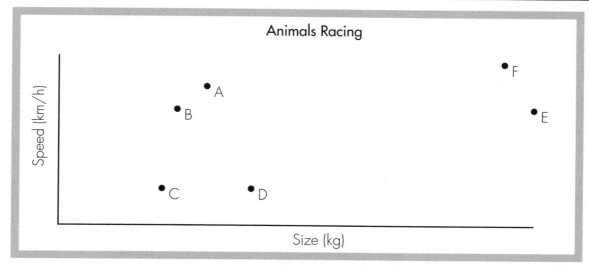

1. List the animals in order from A to F.

   A. _____   B. _____

   C. _____   D. _____

   E. _____   F. _____

2. Explain your list. _____

   _____

   _____

   _____

   _____

Representations

6

Ups and Downs

Drawing and interpreting graphs representing different rates of change

AIM

Students will draw and interpret graphs representing different rates of change.

MATERIALS

- 1 copy of the blackline master (opposite) for each student

REFLECTION

Refer to Graph A in Question 1 on the blackline master. Discuss the trends in the graph: the steeper the graph the faster the speed; if the graph is horizontal there is no movement; and the greater the distance from the ground, the higher the point on the graph.

1 On the board, draw the x and y axes of a graph and label them as shown below.

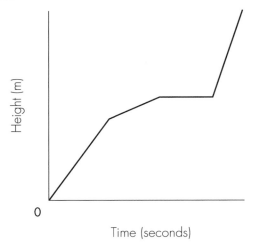

Say, *This graph shows the growth of a new tree over one year. Where on the graph was the tree planted? What part of the graph represents the winter months?* (The horizontal part.) *Why?* (Because plants are often dormant in winter.) *What does the steep part at the top represent?* (Spring.) *Why does it look like this?* (Because plants grow faster in spring.)

2 Read Question 1 on the blackline master with the class. Ask, *What part of the graph represents the first slope? Does the rollercoaster car stop at the top? Does it climb up faster than it goes down?* Ask the students to label each part of their graph as "climbing", "at the top", and "coming down". Ask the students to complete the question. Call on volunteers to share and justify their answers.

3 Read Question 2 with the class. Ask, *How are the night hours represented?* (By the horizontal part at the bottom of the graph.) *What time of day does the top of the graph represent?* (12 noon.) Direct the students to complete the question. Allow time for them to share their answers.

Ups and Downs

Name _____

1. A rollercoaster car starts at the bottom, climbs the first slope slowly, gets to the top and then goes down quickly. Look at the 2 graphs and answer the questions below.

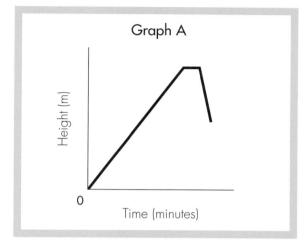

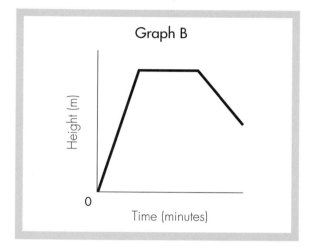

a. Which graph matches the description above? _____

b. Explain why. _____

_____

2. Look at the graphs below.

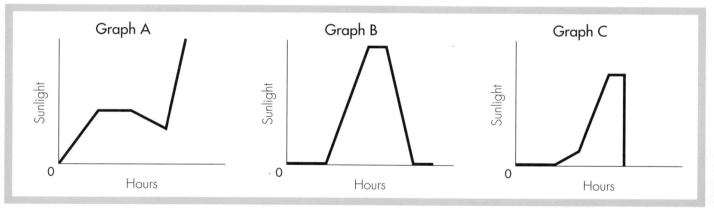

a. Which graph represents the amount of sunlight over 24 hours of a fine day?

_____

b. Explain why. _____

_____

Representations

7

ANSWERS

Equivalence and Equations 1
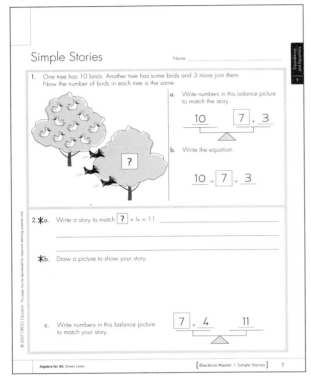

Simple Stories
Name _____

1. One tree has 10 birds. Another tree has some birds and 3 more join them.
Now the number of birds in each tree is the same.

a. Write numbers in this balance picture to match the story.

10 | 7 + 3

b. Write the equation.

10 = 7 + 3

2. *a. Write a story to match ? + 4 = 11. _____

*b. Draw a picture to show your story.

c. Write numbers in this balance picture to match your story.

7 + 4 | 11

Equivalence and Equations 2
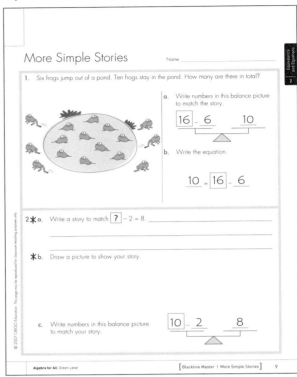

More Simple Stories
Name _____

1. Six frogs jump out of a pond. Ten frogs stay in the pond. How many are there in total?

a. Write numbers in this balance picture to match the story.

16 - 6 | 10

b. Write the equation.

10 = 16 - 6

2. *a. Write a story to match ? - 2 = 8. _____

*b. Draw a picture to show your story.

c. Write numbers in this balance picture to match your story.

10 - 2 | 8

Equivalence and Equations 3
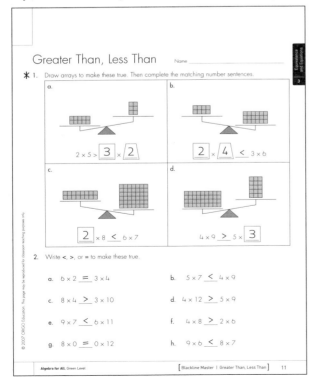

Greater Than, Less Than
Name _____

* 1. Draw arrays to make these true. Then complete the matching number sentences.

a. $2 \times 5 > 3 \times 2$

b. $2 \times 4 < 3 \times 6$

c. $2 \times 8 < 6 \times 7$

d. $4 \times 9 > 5 \times 3$

2. Write <, >, or = to make these true.

a. $6 \times 2 = 3 \times 4$

b. $5 \times 7 < 4 \times 9$

c. $8 \times 4 > 3 \times 10$

d. $4 \times 12 > 5 \times 9$

e. $9 \times 7 < 6 \times 11$

f. $4 \times 8 > 2 \times 6$

g. $8 \times 0 = 0 \times 12$

h. $9 \times 6 < 8 \times 7$

Equivalence and Equations 4
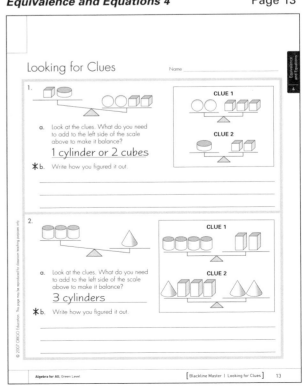

Looking for Clues
Name _____

1.
CLUE 1
CLUE 2

a. Look at the clues. What do you need to add to the left side of the scale above to make it balance?

1 cylinder or 2 cubes

*b. Write how you figured it out.

2.
CLUE 1
CLUE 2

a. Look at the clues. What do you need to add to the left side of the scale above to make it balance?

3 cylinders

*b. Write how you figured it out.

* Answers will vary. This is one example.

Equivalence and Equations 5 — Page 15

Let's Go Shopping   Name _____

Write how many units each item weighs.

1.

Beans = __8__ units   Milk = __5__ units   Popcorn = __3__ units

2.

Coffee = __5__ units   Cheese = __9__ units   Raisins = __7__ units

3.

Mustard = __2__ units   Cream = __5__ units   Mayonnaise = __8__ units

Algebra for All, Green Level   [Blackline Master | Let's Go Shopping]   15

Equivalence and Equations 6 — Page 17

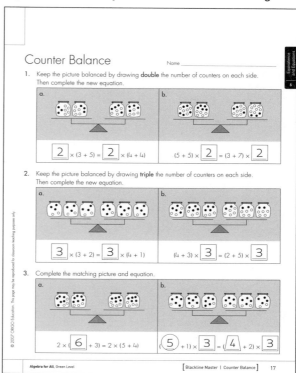

Counter Balance   Name _____

1. Keep the picture balanced by drawing **double** the number of counters on each side. Then complete the new equation.

a. $\boxed{2} \times (3 + 5) = \boxed{2} \times (4 + 4)$ b. $(5 + 5) \times \boxed{2} = (3 + 7) \times \boxed{2}$

2. Keep the picture balanced by drawing **triple** the number of counters on each side. Then complete the new equation.

a. $\boxed{3} \times (3 + 2) = \boxed{3} \times (4 + 1)$ b. $(4 + 3) \times \boxed{3} = (2 + 5) \times \boxed{3}$

3. Complete the matching picture and equation.

a. $2 \times (\boxed{6} + 3) = 2 \times (5 + 4)$ b. $(\boxed{5} + 1) \times \boxed{3} = (\boxed{4} + 2) \times \boxed{3}$

Algebra for All, Green Level   [Blackline Master | Counter Balance]   17

Equivalence and Equations 7 — Page 19

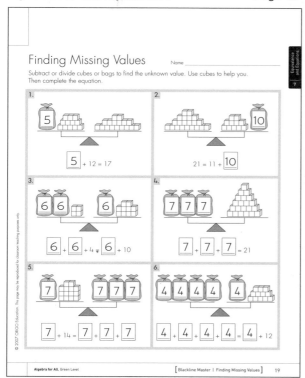

Finding Missing Values   Name _____

Subtract or divide cubes or bags to find the unknown value. Use cubes to help you. Then complete the equation.

1. $\boxed{5} + 12 = 17$

2. $21 = 11 + \boxed{10}$

3. $\boxed{6} + \boxed{6} + 4 = \boxed{6} + 10$

4. $\boxed{7} + \boxed{7} + \boxed{7} = 21$

5. $\boxed{7} + 14 = \boxed{7} + \boxed{7} + \boxed{7}$

6. $\boxed{4} + \boxed{4} + \boxed{4} + \boxed{4} = \boxed{4} + 12$

Algebra for All, Green Level   [Blackline Master | Finding Missing Values]   19

Equivalence and Equations 8 — Page 21

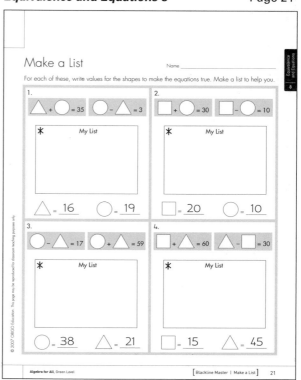

Make a List   Name _____

For each of these, write values for the shapes to make the equations true. Make a list to help you.

1. $\triangle + \bigcirc = 35$ $\bigcirc - \triangle = 3$

   ∗ My List

   $\triangle = $ __16__ $\bigcirc = $ __19__

2. $\square + \bigcirc = 30$ $\square - \bigcirc = 10$

   ∗ My List

   $\square = $ __20__ $\bigcirc = $ __10__

3. $\bigcirc - \triangle = 17$ $\bigcirc + \triangle = 59$

   ∗ My List

   $\bigcirc = $ __38__ $\triangle = $ __21__

4. $\square + \triangle = 60$ $\triangle - \square = 30$

   ∗ My List

   $\square = $ __15__ $\triangle = $ __45__

Algebra for All, Green Level   [Blackline Master | Make a List]   21

∗ Answers will vary. This is one example.

ANSWERS

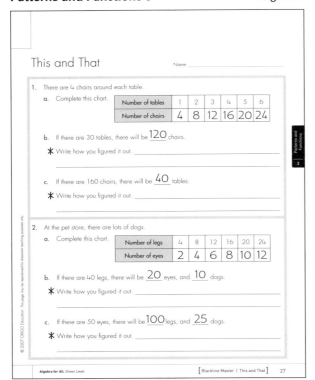

Patterns and Functions 1 — Page 23

Perfect Punch   Name _____

1. To make one bowl of tropical punch, mix 5 cups of orange juice with 1 cup of cranberry juice. Complete this table.

Number of bowls (repeats)	Cups of orange juice	Cups of cranberry juice
1	5	1
2	10	2
3	15	3
4	20	4
5	25	5
10	50	10
12	60	12

2. To make one bowl of berry punch, mix 4 cups of raspberry juice with 2 cups of grape juice.

   a. Use R for raspberry and G for grape. Write 3 parts in the repeating pattern for this punch.

   RRRRGG   RRRRGG   RRRRGG

   b. Complete this table.

Number of bowls (repeats)	Cups of raspberry juice	Cups of grape juice
1	4	2
2	8	4
3	12	6
4	16	8
5	20	10
10	40	20
16	64	32

Algebra for All, Green Level   [Blackline Master | Perfect Punch]   23

Patterns and Functions 2 — Page 25

In the Kitchen   Name _____

1. One batch of muffins needs 1 cup of sugar, 2 cups of butter, and 4 cups of flour. Complete this table.

Number of batches (repeats)	Cups of sugar	Cups of butter	Cups of flour
1	1	2	4
2	2	4	8
3	3	6	12
4	4	8	16
5	5	10	20
10	10	20	40
16	16	32	64

2. One batch of scones needs 3 cups of flour, 2 tablespoons of butter, and 1 cup of milk.

   a. Use the first letter of each ingredient and write 3 parts in the repeating pattern.

   FFFBBM   FFFBBM   FFFBBM

   b. Complete the table below.

Number of batches (repeats)	Cups of flour	Tablespoons of butter	Cups of milk
1	3	2	1
2	6	4	2
3	9	6	3
4	12	8	4
5	15	10	5
10	30	20	10
15	45	30	15

Algebra for All, Green Level   [Blackline Master | In the Kitchen]   25

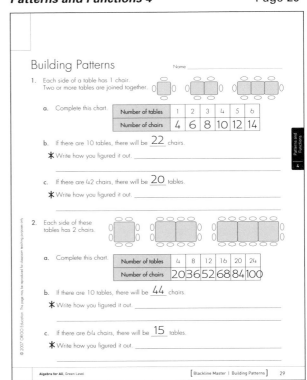

Patterns and Functions 3 — Page 27

This and That   Name _____

1. There are 4 chairs around each table.

   a. Complete this chart.

Number of tables	1	2	3	4	5	6
Number of chairs	4	8	12	16	20	24

   b. If there are 30 tables, there will be **120** chairs.

   ✱ Write how you figured it out. _____

   c. If there are 160 chairs, there will be **40** tables.

   ✱ Write how you figured it out. _____

2. At the pet store, there are lots of dogs.

   a. Complete this chart.

Number of legs	4	8	12	16	20	24
Number of eyes	2	4	6	8	10	12

   b. If there are 40 legs, there will be **20** eyes, and **10** dogs.

   ✱ Write how you figured it out. _____

   c. If there are 50 eyes, there will be **100** legs, and **25** dogs.

   ✱ Write how you figured it out. _____

Algebra for All, Green Level   [Blackline Master | This and That]   27

Patterns and Functions 4 — Page 29

Building Patterns   Name _____

1. Each side of a table has 1 chair. Two or more tables are joined together.

   a. Complete this chart.

Number of tables	1	2	3	4	5	6
Number of chairs	4	6	8	10	12	14

   b. If there are 10 tables, there will be **22** chairs.

   ✱ Write how you figured it out. _____

   c. If there are 42 chairs, there will be **20** tables.

   ✱ Write how you figured it out. _____

2. Each side of these tables has 2 chairs.

   a. Complete this chart.

Number of tables	4	8	12	16	20	24
Number of chairs	20	36	52	68	84	100

   b. If there are 10 tables, there will be **44** chairs.

   ✱ Write how you figured it out. _____

   c. If there are 64 chairs, there will be **15** tables.

   ✱ Write how you figured it out. _____

Algebra for All, Green Level   [Blackline Master | Building Patterns]   29

✱ Answers will vary. This is one example.

Patterns and Functions 5 — Page 31

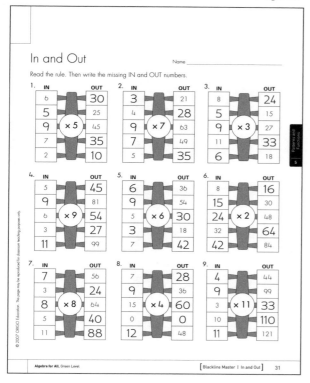

In and Out

Name _____

Read the rule. Then write the missing IN and OUT numbers.

1.

IN	× 5	OUT
6		30
5		25
9		45
7		35
2		10

2.

IN	× 7	OUT
3		21
4		28
9		63
7		49
5		35

3.

IN	× 3	OUT
8		24
5		15
9		27
11		33
6		18

4.

IN	× 9	OUT
5		45
9		81
6		54
3		27
11		99

5.

IN	× 6	OUT
6		36
9		54
5		30
3		18
7		42

6.

IN	× 2	OUT
8		16
15		30
24		48
32		64
42		84

7.

IN	× 8	OUT
7		56
3		24
8		64
5		40
11		88

8.

IN	× 4	OUT
7		28
9		36
15		60
0		0
12		48

9.

IN	× 11	OUT
4		44
9		99
3		33
10		110
11		121

Algebra for All, Green Level   [Blackline Master | In and Out]   31

Patterns and Functions 6 — Page 33

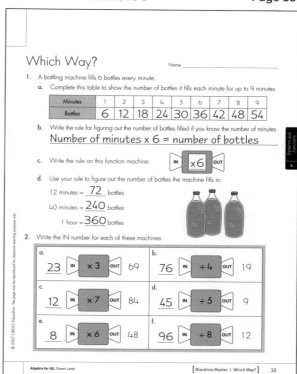

Which Way?

Name _____

1. A bottling machine fills 6 bottles every minute.

   a. Complete this table to show the number of bottles it fills each minute for up to 9 minutes.

Minutes	1	2	3	4	5	6	7	8	9
Bottles	6	12	18	24	30	36	42	48	54

   b. Write the rule for figuring out the number of bottles filled if you know the number of minutes.

   <u>Number of minutes x 6 = number of bottles</u>

   c. Write the rule on this function machine.   IN ×6 OUT

   d. Use your rule to figure out the number of bottles the machine fills in:

   12 minutes = __72__ bottles

   40 minutes = __240__ bottles

   1 hour = __360__ bottles

2. Write the IN number for each of these machines.

   a. __23__ IN ×3 OUT 69

   b. __76__ IN ÷4 OUT 19

   c. __12__ IN ×7 OUT 84

   d. __45__ IN ÷5 OUT 9

   e. __8__ IN ×6 OUT 48

   f. __96__ IN ÷8 OUT 12

Algebra for All, Green Level   [Blackline Master | Which Way?]   33

Patterns and Functions 7 — Page 35

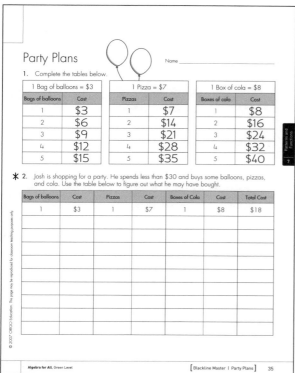

Party Plans

Name _____

1. Complete the tables below.

1 Bag of balloons = $3	
Bags of balloons	Cost
1	$3
2	$6
3	$9
4	$12
5	$15

1 Pizza = $7	
Pizzas	Cost
1	$7
2	$14
3	$21
4	$28
5	$35

1 Box of cola = $8	
Boxes of cola	Cost
1	$8
2	$16
3	$24
4	$32
5	$40

✱ 2. Josh is shopping for a party. He spends less than $30 and buys some balloons, pizzas, and cola. Use the table below to figure out what he may have bought.

Bags of balloons	Cost	Pizzas	Cost	Boxes of Cola	Cost	Total Cost
1	$3	1	$7	1	$8	$18

Algebra for All, Green Level   [Blackline Master | Party Plans]   35

Patterns and Functions 8 — Page 37

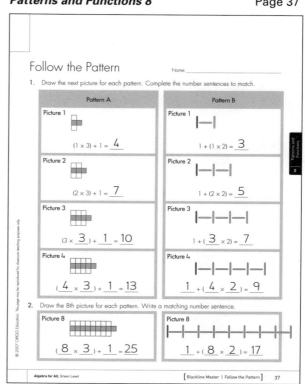

Follow the Pattern

Name _____

1. Draw the next picture for each pattern. Complete the number sentences to match.

Pattern A	Pattern B
Picture 1 $(1 \times 3) + 1 =$ __4__	Picture 1 $1 + (1 \times 2) =$ __3__
Picture 2 $(2 \times 3) + 1 =$ __7__	Picture 2 $1 + (2 \times 2) =$ __5__
Picture 3 $(3 \times$ __3__$) +$ __1__ $=$ __10__	Picture 3 $1 + ($ __3__ $\times 2) =$ __7__
Picture 4 __4__ $\times 3) +$ __1__ $=$ __13__	Picture 4 __1__ $+ ($ __4__ $\times 2) =$ __9__

2. Draw the 8th picture for each pattern. Write a matching number sentence.

| Picture 8 __8__ $\times 3) +$ __1__ $=$ __25__ | Picture 8 __1__ $+ ($ __8__ $\times 2) =$ __17__ |

Algebra for All, Green Level   [Blackline Master | Follow the Pattern]   37

✱ Answers will vary. This is one example.

ANSWERS

For Sale

Name _____

Muffins are sold in packs of 3, cookies are sold in packs of 6, and cakes are sold in packs of 5.

1. Complete the tables below.

Muffins	Packs of 3
12	4
45	15
66	22
123	41
150	50

Cookies	Packs of 6
12	2
36	6
126	21
108	18
312	52

Cakes	Packs of 5
10	2
85	17
125	25
90	18
200	40

2. Cookies cost $1.50 per pack. Complete this table to show the total dollar value for each number of cookies.

Cookies	Packs	Cost per pack	Total dollar value
36	6	$1.50	$9.00
72	12	$1.50	$18.00
108	18	$1.50	$27.00
120	20	$1.50	$30.00
318	53	$1.50	$79.50

3. Muffins cost $1.20 per pack. Complete this table to show the total dollar value for each number of muffins.

Muffins	Packs of 3	Cost per pack	Total dollar value
36	12	$1.20	$14.40
81	27	$1.20	$32.40
90	30	$1.20	$36.00
120	40	$1.20	$48.00
180	60	$1.20	$72.00

Buying and Selling

Name _____

1. Burgers cost $6 each, pizzas cost $7 each, and salads cost $8 each. Complete the tables below.

Burgers	Cost
4	$24
14	$84
13	$78
10	$60
8	$48

Pizzas	Cost
8	$56
11	$77
15	$105
9	$63
7	$49

Salads	Cost
7	$56
20	$160
13	$104
14	$112
5	$40

2. Muffins are sold in packs of 3, cookies are sold in packs of 6, and cakes are sold in packs of 2. Complete the tables below.

Muffins	Packs of 3
12	4
36	12
48	16
69	23
123	41

Cookies	Packs of 6
24	4
72	12
96	16
84	14
48	8

Cakes	Packs of 2
122	61
96	48
56	28
240	120
36	18

3. Cookies are sold in packs of 6 and cost $1.50 per pack. Complete the table below.

Cookies	Packs of 6	Cost per pack	Total dollar value
24	4	$1.50	$6.00
60	10	$1.50	$15.00
42	7	$1.50	$10.50
54	9	$1.50	$13.50
48	8	$1.50	$12.00

Go Grids

Name _____

1. Write 2 turnaround multiplication facts that can be used to figure out the area of each grid.

a.
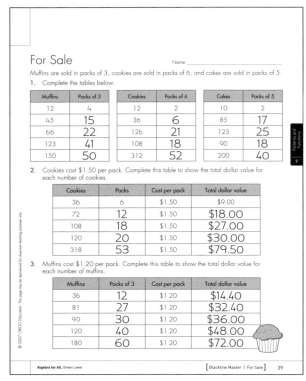
$3 \times \underline{6} = \underline{18}$ $\underline{6} \times 3 = \underline{18}$

b.
$4 \times \underline{7} = \underline{28}$ $\underline{7} \times \underline{4} = \underline{28}$

c.
$5 \times \underline{4} = \underline{20}$ $\underline{4} \times 5 = \underline{20}$

d.
$\underline{2} \times \underline{8} = \underline{16}$ $\underline{8} \times \underline{2} = \underline{16}$

2. Draw grids to show each expression. Then write 2 turnaround multiplication facts to match.

a. 7×3
$\underline{7} \times \underline{3} = \underline{21}$ $\underline{3} \times \underline{7} = \underline{21}$

b. 9×2
$\underline{9} \times \underline{2} = \underline{18}$ $\underline{2} \times \underline{9} = \underline{18}$

c. 6×5
$\underline{6} \times \underline{5} = \underline{30}$ $\underline{5} \times \underline{6} = \underline{30}$

d. 4×6
$\underline{4} \times \underline{6} = \underline{24}$ $\underline{6} \times \underline{4} = \underline{24}$

Does it Commute?

Name _____

1. Write the answer to each equation and its turnaround. Draw jumps to help you.

a.
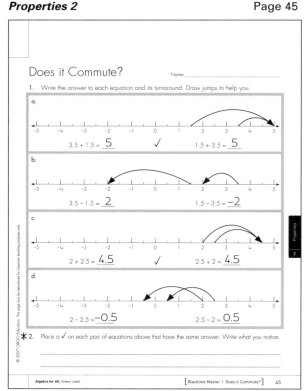
$3.5 + 1.5 = \underline{5}$ ✓ $1.5 + 3.5 = \underline{5}$

b.
$3.5 - 1.5 = \underline{2}$ $1.5 - 3.5 = \underline{-2}$

c.
$2 + 2.5 = \underline{4.5}$ ✓ $2.5 + 2 = \underline{4.5}$

d.
$2 - 2.5 = \underline{-0.5}$ $2.5 - 2 = \underline{0.5}$

✱2. Place a ✓ on each pair of equations above that have the same answer. Write what you notice.

✱ Answers will vary. This is one example.

Properties 3 — Page 47

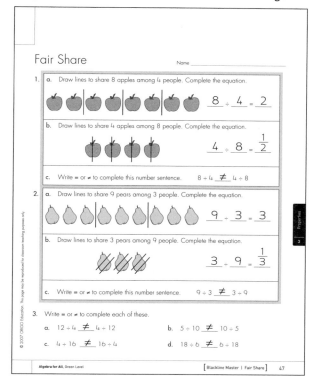

Fair Share
Name _____

1. a. Draw lines to share 8 apples among 4 people. Complete the equation.

   $8 ÷ 4 = 2$

   b. Draw lines to share 4 apples among 8 people. Complete the equation.

   $4 ÷ 8 = \dfrac{1}{2}$

   c. Write = or ≠ to complete this number sentence. $8 ÷ 4 ≠ 4 ÷ 8$

2. a. Draw lines to share 9 pears among 3 people. Complete the equation.

   $9 ÷ 3 = 3$

   b. Draw lines to share 3 pears among 9 people. Complete the equation.

   $3 ÷ 9 = \dfrac{1}{3}$

   c. Write = or ≠ to complete this number sentence. $9 ÷ 3 ≠ 3 ÷ 9$

3. Write = or ≠ to complete each of these.

   a. $12 ÷ 4 ≠ 4 ÷ 12$ b. $5 ÷ 10 ≠ 10 ÷ 5$

   c. $4 ÷ 16 ≠ 16 ÷ 4$ d. $18 ÷ 6 ≠ 6 ÷ 18$

Algebra for All, Green Level   [Blackline Master | Fair Share] 47

Properties 4 — Page 49

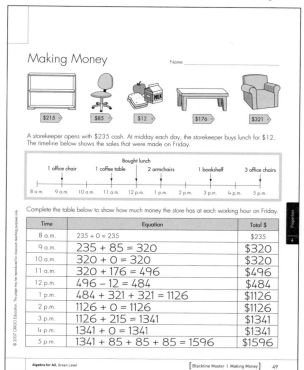

Making Money
Name _____

$215 $85 $12 $176 $321

A storekeeper opens with $235 cash. At midday each day, the storekeeper buys lunch for $12. The timeline below shows the sales that were made on Friday.

Bought lunch

1 office chair   1 coffee table   2 armchairs   1 bookshelf   3 office chairs

8 a.m. 9 a.m. 10 a.m. 11 a.m. 12 p.m. 1 p.m. 2 p.m. 3 p.m. 4 p.m. 5 p.m.

Complete the table below to show how much money the store has at each working hour on Friday.

Time	Equation	Total $
8 a.m.	235 + 0 = 235	$235
9 a.m.	235 + 85 = 320	$320
10 a.m.	320 + 0 = 320	$320
11 a.m.	320 + 176 = 496	$496
12 p.m.	496 − 12 = 484	$484
1 p.m.	484 + 321 + 321 = 1126	$1126
2 p.m.	1126 + 0 = 1126	$1126
3 p.m.	1126 + 215 = 1341	$1341
4 p.m.	1341 + 0 = 1341	$1341
5 p.m.	1341 + 85 + 85 + 85 = 1596	$1596

Algebra for All, Green Level   [Blackline Master | Making Money] 49

Properties 5 — Page 51

Buying and Sharing
Name _____

1. Haley buys 9 bunches of 8 flowers.

   a. Complete a matching equation.

   Haley's flowers = $9 × 8$

   b. Haley shares her flowers equally with her 3 friends. Complete the equation to show Haley's share of flowers.

   Haley's share = $9 × 8 ÷ 4$

   ✱ c. Write Haley's share and how you figured it out.

   Hayley's share = 18 flowers

2. Anita saves $15 each week for 4 weeks.

   a. Complete a matching equation.

   Anita's savings = $15 × 4$

   b. Anita shares her money equally with her 2 sisters. Complete the equation to show Anita's share of money.

   Anita's share = $15 × 4 ÷ 3$

   ✱ c. Write Anita's share and how you figured it out.

   Anita's share = $20

Algebra for All, Green Level   [Blackline Master | Buying and Sharing] 51

Representations 1 — Page 53

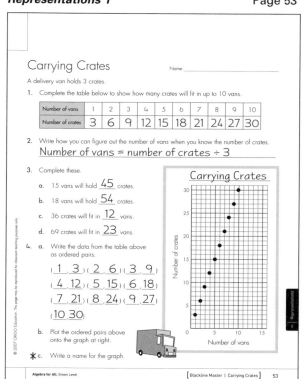

Carrying Crates
Name _____

A delivery van holds 3 crates.

1. Complete the table below to show how many crates will fit in up to 10 vans.

Number of vans	1	2	3	4	5	6	7	8	9	10
Number of crates	3	6	9	12	15	18	21	24	27	30

2. Write how you can figure out the number of vans when you know the number of crates.

   Number of vans = number of crates ÷ 3

3. Complete these.

   a. 15 vans will hold 45 crates.

   b. 18 vans will hold 54 crates.

   c. 36 crates will fit in 12 vans.

   d. 69 crates will fit in 23 vans.

4. a. Write the data from the table above as ordered pairs.

   (1 , 3) (2 , 6) (3 , 9)
   (4 , 12) (5 , 15) (6 , 18)
   (7 , 21) (8 , 24) (9 , 27)
   (10 , 30)

   b. Plot the ordered pairs above onto the graph at right.

   ✱ c. Write a name for the graph.

   Carrying Crates

Algebra for All, Green Level   [Blackline Master | Carrying Crates] 53

✱ Answers will vary. This is one example.

ANSWERS

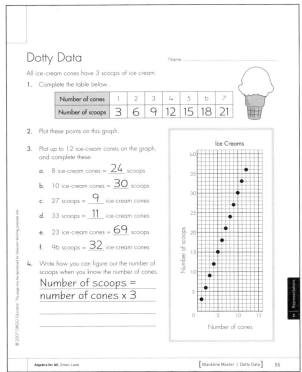

Dotty Data   Name _____

All ice-cream cones have 3 scoops of ice cream.

1. Complete the table below.

Number of cones	1	2	3	4	5	6	7
Number of scoops	3	6	9	12	15	18	21

2. Plot these points on this graph.

3. Plot up to 12 ice-cream cones on the graph, and complete these.

   a. 8 ice-cream cones = **24** scoops

   b. 10 ice-cream cones = **30** scoops

   c. 27 scoops = **9** ice-cream cones

   d. 33 scoops = **11** ice-cream cones

   e. 23 ice-cream cones = **69** scoops

   f. 96 scoops = **32** ice-cream cones

4. Write how you can figure out the number of scoops when you know the number of cones.

   Number of scoops = number of cones x 3

Ice Creams

Algebra for All, Green Level   [Blackline Master I Dotty Data]   55

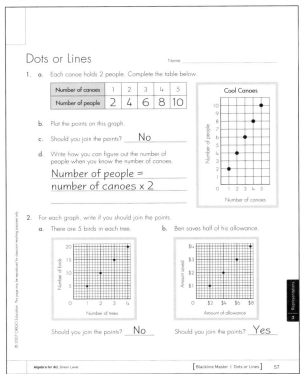

Dots or Lines   Name _____

1. a. Each canoe holds 2 people. Complete the table below.

Number of canoes	1	2	3	4	5
Number of people	2	4	6	8	10

   b. Plot the points on this graph.

   c. Should you join the points? **No**

   d. Write how you can figure out the number of people when you know the number of canoes.

   Number of people = number of canoes x 2

Cool Canoes

2. For each graph, write if you should join the points.

   a. There are 5 birds in each tree.

   b. Ben saves half of his allowance.

Should you join the points? **No**   Should you join the points? **Yes**

Algebra for All, Green Level   [Blackline Master I Dots or Lines]   57

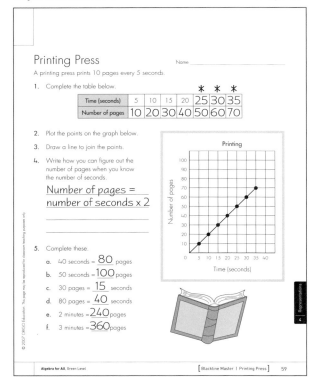

Printing Press   Name _____

A printing press prints 10 pages every 5 seconds.

1. Complete the table below.

                             * * *

Time (seconds)	5	10	15	20	25	30	35
Number of pages	10	20	30	40	50	60	70

2. Plot the points on the graph below.

3. Draw a line to join the points.

4. Write how you can figure out the number of pages when you know the number of seconds.

   Number of pages = number of seconds x 2

Printing

5. Complete these.

   a. 40 seconds = **80** pages

   b. 50 seconds = **100** pages

   c. 30 pages = **15** seconds

   d. 80 pages = **40** seconds

   e. 2 minutes = **240** pages

   f. 3 minutes = **360** pages

Algebra for All, Green Level   [Blackline Master I Printing Press]   59

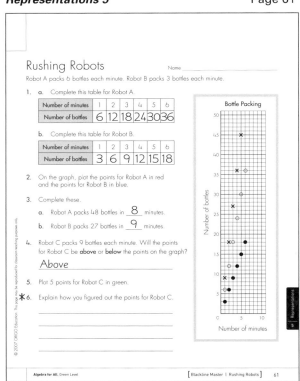

Rushing Robots   Name _____

Robot A packs 6 bottles each minute. Robot B packs 3 bottles each minute.

1. a. Complete this table for Robot A.

Number of minutes	1	2	3	4	5	6
Number of bottles	6	12	18	24	30	36

   b. Complete this table for Robot B.

Number of minutes	1	2	3	4	5	6
Number of bottles	3	6	9	12	15	18

2. On the graph, plot the points for Robot A in red and the points for Robot B in blue.

3. Complete these.

   a. Robot A packs 48 bottles in **8** minutes.

   b. Robot B packs 27 bottles in **9** minutes.

4. Robot C packs 9 bottles each minute. Will the points for Robot C be **above** or **below** the points on the graph?

   Above

5. Plot 5 points for Robot C in green.

*6. Explain how you figured out the points for Robot C.

Bottle Packing

Algebra for All, Green Level   [Blackline Master I Rushing Robots]   61

* Answers will vary. This is one example.

Puzzle Graphs

Name _____

This table shows the average speed of 6 animals.
Each animal in the table is represented
on the scatter plot below.

Animal	Average speed
Cat	50 km/h
Chicken	17 km/h
Grizzly bear	50 km/h
Hunting dog	62 km/h
Lion	72 km/h
Pig	17 km/h

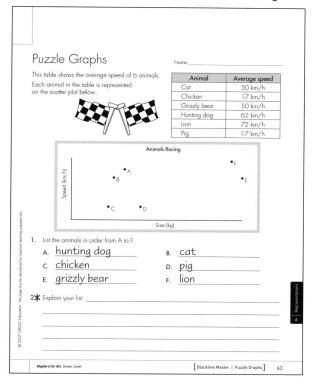

Animals Racing

1. List the animals in order from A to F.

    A. hunting dog   B. cat

    C. chicken   D. pig

    E. grizzly bear   F. lion

2✱ Explain your list. _____
_____
_____
_____

Ups and Downs

Name _____

1. A rollercoaster car starts at the bottom, climbs the first slope slowly, gets to the top
    and then goes down quickly. Look at the 2 graphs and answer the questions below.

    Graph A
    at the top
    climbing / coming down
    Height (m)
    0   Time (minutes)

    Graph B
    Height (m)
    0   Time (minutes)

    a. Which graph matches the description above? Graph A

    ✱ b. Explain why. _____
    _____

2. Look at the graphs below.

    Graph A   Graph B   Graph C
    Sunlight   Sunlight   Sunlight
    0   Hours   0   Hours   0   Hours

    a. Which graph represents the amount of sunlight over 24 hours of a fine day?
        Graph B

    ✱ b. Explain why. _____
    _____

✱ Answers will vary. This is one example.

Assessment Summary

Name _____

	Lesson	Page	A	B	C	D	Date
Equivalence and Equations	Simple Stories	6					
	More Simple Stories	8					
	Greater Than, Less Than	10					
	Looking for Clues	12					
	Let's Go Shopping	14					
	Counter Balance	16					
	Finding Missing Values	18					
	Make a List	20					
Patterns and Functions	Perfect Punch	22					
	In the Kitchen	24					
	This and That	26					
	Building Patterns	28					
	In and Out	30					
	Which Way?	32					
	Party Plans	34					
	Follow the Pattern	36					
	For Sale	38					
	Buying and Selling	40					
Properties	Go Grids	42					
	Does it Commute?	44					
	Fair Share	46					
	Making Money	48					
	Buying and Sharing	50					
Representations	Carrying Crates	52					
	Dotty Data	54					
	Dots or Lines	56					
	Printing Press	58					
	Rushing Robots	60					
	Puzzle Graphs	62					
	Ups and Downs	64					